Y0-EAO-980

BASIC
MATHEMATICAL SKILLS

BASIC MATHEMATICAL SKILLS

THOMAS G. SMITHSI

Dean, Preparatory Department
Technical Career Institutes

PRENTICE-HALL, INC.
Englewood Cliffs, New Jersey

Library of Congress Cataloging in Publication Data

Smithsi, Thomas G
Basic mathematical skills.

1. Arithmetic–1961- I. Title.
QA107.S56 513 73-18462
ISBN 0-13-063420-4

Printed in the United States of America

10 9 8 7 6 5 4 3

PRENTICE-HALL INTERNATIONAL, INC., London
PRENTICE-HALL OF AUSTRALIA, PTY. LTD., Sydney
PRENTICE-HALL OF CANADA, LTD., Toronto
PRENTICE-HALL OF INDIA PRIVATE LIMITED, New Delhi
PRENTICE-HALL OF JAPAN, INC., Tokyo

To my wife Roza
and
my two daughters Billie and Bunny

CONTENTS

WHAT? . . . . ANOTHER BOOK ON ARITHMETIC?

For as long as the 3 R's–Reading, 'Riting, and 'Rithmetic–remain considered as the necessary building blocks of knowledge a human being must master in order to survive in this society of ours, we as educators, are obligated to constantly research, experiment, and develop methods and techniques that will make skill in these areas accessible to all. This book is my effort and contribution to that end.

I do not claim to have the "Fountain of Knowledge"–all I know is what the content of this book has done for my students, who represent a cross-section of what is found in the labor market and schools of various levels.

The courses taught at RCA Institutes require the student to be thoroughly knowledgeable and skilled in arithmetic before any other math topic can be considered. As Senior Instructor of the Preparatory Department, one of my duties is to prepare remedial arithmetic programs, but finding a textbook to supplement our work can be difficult. Some of the publications emphasize the "modern math" approach. As a teacher, I personally find them to be very interesting, but they do not appeal to my students because many of them are not academically oriented–some have poor study habits, while others have reading problems. Other publications use the "programmed" technique. The idea of presenting a topic in small bits and gradually working it up to a climax is good–however, the books I received were but a weak attempt at the philosophy and goals of programming.

To overcome the disadvantages of currently available publications, I proceeded to compile what I thought to be the good points of both modern math and of programming into a SEMI-PROGRAMMED WORKTEXT that would appeal to the student as well as the teacher.

In my opinion, the key to all learning is the teacher. No amount of printed words can substitute for the experience, wisdom, warmth, understanding, encouragement, and praise a student constantly needs to be motivated. This book was written to make the task of teaching these basic mathematical skills a little easier by:

1) presenting the text on an uncrowded one- or two-column page that is easier to read and understand than large blocks of words;

2) developing operations, first intuitively and then by sketches, numberline applications, etc., developing a method of solution and then, by simplification to the traditional method, making an important link between modern math and traditional;

3) using an informal conversational tone, guiding the student, pointing out the common difficulties, possible error situations, important observations, and things he must remember;

4) using training aids for all operations developed to provide a good background for future equation solving;

5) using estimates to develop arithmetical common sense; and

6) using monetary problems and guessing games to develop intuition.

This should offer the teacher many possibilities as to how to set up a learning situation including

a) student aid on an individual basis,

b) student aid on a grouping basis, and

c) partial presentation, leaving the remainder to the student.

We are aware that many remedial students do not need to review the entire arithmetic section. To help the teacher make a decision, we have preceded each chapter with a PRE-TEST. If the student can handle the problems, he may omit the chapter. All teachers are aware that the better students often have a rudimentary knowledge of the operations, but could still benefit from reading through the entire chapter. This decision lies with the teacher alone.

The work area consists of 4,200 problems in the form of:

STEP-BY-STEP TEST with prescribed time limits.

Students judge only if they can do a problem—they must be impressed that failure is likely if a test containing many problems is not completed.

ACHIEVEMENT TESTS to prove the before and after results.

REFRESHER TESTS to maintain the acquired skills.

RELATED WORD PROBLEMS.

A FINAL GENERAL MATHEMATICS TEST comparable to those assembled by couseling and research organizations for the benefit of institutions and industry, to evaluate the arithmetic knowledge and skills of the individual for proper placement.

THOMAS G. SMITHSI
RCA Institutes

BASIC
MATHEMATICAL SKILLS

Chapter 1

OUR NUMBERING SYSTEM; ADDITION

PRETEST–ADDITION

This may sound crazy, but errors in addition are more often due to a faulty mastery of the basic facts (addition tables) than any other cause.

The only procedure that will improve both accuracy and speed in addition is complete mastery of these basic facts.

A time limit for writing the answers for the addition tests has been set to prevent you from counting on your fingers, tapping with the pencil, or using any other roundabout method of arriving at the answers.

To diagnose your weakness correctly we must adopt a step-by-step corrective and remedial method to bring about an improvement.

Take the pretest.

If you get 20 or more correct answers you may omit the chapter on addition.

However–

if your right answers are fewer than the number above, it could be because of one of these reasons:

1) You hesitated a lot when you added and lost count.

2) You did not write down the numbers you were to carry.

3) You did not check each column as you added.

4) You counted by ones and worked too slowly.

5) You did not make sensible estimates of the sums, as a check.

If your problem exists in any of the stated areas–do yourself a favor–and go through the chapter on addition.

We're sure it will do you a lot of good.

PRETEST–ADDITION

Add 8 to each of the following numbers: 13___ 24___ 36___ 18___ 27___ 49___ 29___ 35___.

Find the sum of 15 and 7 added to 15_____. Find the sum of 24 and 8 added to 19_____.

Add:

9	86	64	7,068	8,040	8,379	64,964,663	4,985
5	77	87	9,098	3,090	6,429	83,355,854	4,567
7	97	98			3,994	18,651,748	9,325
4	69	55	396	3,096	2,547	18,656,427	5,776
8	38	86	606	6,206			4,668
							7,979
							3,856
							8,379

8	57	357	9,454
7	49	947	4,587
9	78	866	9,186
4	63	598	7,575
6	94	467	5,839
9	55	764	4,684
8	86	685	4,589
5	97	398	6,372

What number added to 8 and 7 will equal 26?_____

What number added to 12 and 6 will equal 30?_____

What number added to 23 and 9 will equal 40?_____

12 + 7 + 13 + 8 + 14 = _____

8 + 9 + 7 + 13 + 22 + 6 = _____

OUR NUMBERING SYSTEM

NUMBERS AND NUMERALS

When you think of "how many" objects you possess, your mind develops an "idea" in the form of a NUMBER.
Numbers cannot be seen or written.
Over the years man has used SYMBOLS to represent these numbers.
The symbols used are called NUMERALS, because they are used to represent numbers.

Early man used his ten fingers to record the number of things he wanted to count.
Modern man uses ten symbols to record the number of things he wants to count.

The ten symbols are:

0, 1, 2, 3, 4, 5, 6, 7, 8, 9

The Latin word for ten is "decem."
Our numbering system is made up of ten symbols.
Mathematicians borrowed from the Latin and decided to call our numbering system the DECIMAL SYSTEM.

The Latin word for toe or finger is "digitus."
Every finger and toe is commonly referred to as a DIGIT.
Our ten symbols are also referred to as DIGITS.

There will be times when we will use the word NUMBER when we should really use the word NUMERAL.
It is a mistake that has become a common practice.
We know they mean different things.

We think of a NUMBER.
We write it as a NUMERAL.
We write the numeral using DIGITS.

ORDER AMONG THE NUMERALS

Man established ORDER among his chosen symbols by making each symbol represent a given number.

The NUMBER LINE visually illustrates this order.

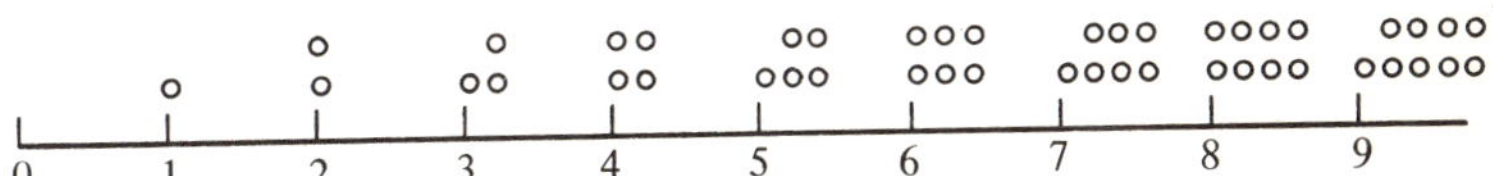

Each symbol corresponds with one and only one point on the number line.
This form of presentation is called a ONE-TO-ONE CORRESPONDENCE.

Each successive point on the number line has a value "one more" than the preceding point.
The numbers increase in value as they move to the right.
Each symbol, except 0, represents a number that is GREATER THAN the symbol on its left.
The farther we go to the right, the greater the number it represents. This means we will always be able to count one more.
Is there an end to the number line?
Of course not; it is INFINITE.

The farther we go to the left the less the number it represents.
Each symbol, including 0, represents a number that is LESS THAN the symbols to its right.

Visually we can understand why 4 is GREATER THAN 3 because 4 is to the right of 3.

We can also understand why 8 is LESS THAN 9 because 8 is to the left of 9.

COMPLETE EACH OF THE FOLLOWING STATEMENTS:

1. A number cannot be seen or written; it is an ______________.
2. Symbols used to represent numbers are called ______________.
3. Our numbering system is made up of ______________ symbols.
4. The symbol 3 can be called a numeral or a ______________.
5. Our numbering system is called the decimal system because it is made up of ______________ symbols.
6. Our numbering system is made up of ______________ digits.
7. The symbols 0, 1, 2, 3, 4, 5, 6, 7, 8 and 9 may be called numerals or ______________.
8. Arranging the ten symbols so that each is meant to represent a definite number is putting these symbols in ______________.
9. To illustrate this order visually, we use the ______________.
10. On the number line, the numbers ______________ in value as they move to the right.
11. 6 is greater than 5 because 6 is to the ______________ of 5.
12. 6 is the successor to ______________.
13. 6 is less than 7 because 6 is to the ______________ of 7.
14. The successor to 6 is ______________.
15. "One-to-one correspondence" means that for every point on the number line there is a definite ______________.

THE ORIGIN OF OUR DECIMAL SYSTEM

With only ten fingers to work with, the largest number of things early man could count was ten. If he wanted to count beyond ten, he had to start all over again.
To record the number of times he started from the beginning, he hired the services of a friend. Each finger of the second set of hands was given the value of TEN.
When and if the fingers of the second set of hands were all used up, he would call in a third set of hands to record the HUNDREDS.

The fingers of each set of hands represented a different set of values.

The fingers of the first set of hands represented the values of ONES.
The fingers of the second set of hands represented the values of TENS.
The fingers of the third set of hands represented the values of HUNDREDS.

PLACE–VALUE SYSTEM

Mathematicians borrowed this idea and developed it into the PLACE-VALUE SYSTEM. The place-value system permits us to use only ten symbols and yet express the largest or the smallest number man can conceive.

When man used up all of his ten fingers, he had to start all over again.
When we use up all of the ten symbols, we, too, will have to start all over again.

CLOCK MATHEMATICS

A clever training aid called CLOCK MATHEMATICS is used to introduce the mechanics of the place-value system. The following are three TEN-HOUR clocks.

The ten numerals painted on each clock represent the ten symbols of our decimal system.
The starting point for each clock is 0.
The position of each clock determines the place value.
With each complete revolution, the hand of each clock comes back to the starting point, which is 0.

When the hand of the ONES clock returns to the starting point, it triggers a "carry" switch on the second clock, causing the hand of the TENS clock to go up one digit, to the numeral 1.

Hundreds

Tens

Ones

With the second complete revolution of the ones hand, the TENS clock will again be triggered to move its hand another digit, this time to the numeral 2.

Hundreds

Tens

Ones

This action repeats itself over and over until the hand of the second clock also returns to the starting point. When that happens, the "carry" switch of the HUNDREDS clock is triggered to move its hand one digit to the numeral 1.

Hundreds

Tens

Ones

What number is represented by the following show of hands?

Third set of hands	Second set of hands	First set of hands	
			1. ______
			2. ______
			3. ______
			4. ______
			5. ______
			6. ______
			7. ______
			8. ______

What number is represented by the following show of clocks?

Third clock	Second clock	First clock	
			9. ______
			10. ______
			11. ______
			12. ______
			13. ______
			14. ______
			15. ______
			16. ______

THE IMPORTANCE OF 10 IN OUR DECIMAL SYSTEM

Man modified the clock arrangement into a COLUMN arrangement.

Each column is given a definite name.
Its name depends upon its position (place).

PLACE VALUES IN THE DECIMAL SYSTEM

COLUMN NUMBER											
12	11	10	9	8	7	6	5	4	3	2	1
Hundred BILLIONS	Ten BILLIONS	BILLIONS	Hundred MILLIONS	Ten MILLIONS	MILLIONS	Hundred THOUSANDS	Ten THOUSANDS	THOUSANDS	HUNDREDS	TENS	ONES

Each column is given a value that is TEN TIMES as large as the column on its right.

10	ONES	=	1	TEN
10	TENS	=	1	HUNDRED
10	HUNDREDS	=	1	THOUSAND
10	THOUSANDS	=	1	TEN THOUSAND
10	TEN THOUSANDS	=	1	HUNDRED THOUSAND
10	HUNDRED THOUSANDS	=	1	MILLION
10	MILLIONS	=	1	TEN MILLION
10	TEN MILLIONS	=	1	HUNDRED MILLION
10	HUNDRED MILLIONS	=	1	BILLION
10	BILLIONS	=	1	TEN BILLION
10	TEN BILLIONS	=	1	HUNDRED BILLION

WHOLE NUMBERS REPRESENTED BY DIGITS

In the numeral 2,563

The place value of 2 is	1,000
The place value of 5 is	100
The place value of 6 is	10
The place value of 3 is	1

The value of a digit in a numeral is the PRODUCT (multiplication) of the digit times the place value of the digit.

In the numeral 2,563

The value of 2 is	2	x	1,000	=	2,000
The value of 5 is	5	x	100	=	500
The value of 6 is	6	x	10	=	60
The value of 3 is	3	x	1	=	3

The value of the number represented by the numeral 2,563 is the SUM (addition) of the values represented by the digits in the numeral.

In the numeral 2,563

```
2,000
  500
   60
    3
-----
2,563
```

In the numeral 170, the zero means NO ONES.
In the numeral 305, the zero means NO TENS.
In the numeral 4,075, the zero means NO HUNDREDS.
In the numeral 60,804, the zeros mean NO TENS and NO THOUSANDS.

The zero in these cases is used as a PLACEHOLDER. Without it, it would be impossible to write a numeral for numbers larger than 9.

COMPLETE EACH OF THE FOLLOWING STATEMENTS:

1. In our place-value system each column has a value that is __10__ times greater than the value of the column on its right.

2. As we move to the left, the value of each column increases by __10__ times.

3. The zero (0) in the numeral 10 is used as a placeholder because it is used to "hold" the __10__ place.

4. In the numeral 1,000 there are __3__ placeholders.

5. In the numeral 7,055 a zero placeholder is located in the __100__ place.

6. The place that a digit occupies in a numeral determines its ________ in the numeral.

7. The value of any digit in a numeral depends upon two things, the digit itself and its ________ in the numeral.

8. Assigning a name to each column is called the ____________ system.

9. When a numeral does not have a value for a given place, the digit 0 is used as a __________.

10. The largest number of ones we can write in the ones place is __9__.

11. The largest number of tens we can write in the tens place is __99__.

12. The largest number of hundreds we can write in the hundreds place is __999__.

13. In the numeral 624

    The 6 represents __600__.
    The 2 represents __20__.
    The 4 represents __4__.

14. In the numeral 2,063

    The 2 represents __2000__.
    The 0 represents __0__.
    The 6 represents __60__.
    The 3 represents __3__.

15. Write a numeral with a 3 in the tens place and a 5 in the ones place __35__.

16. Write a numeral with a 3 in the ones place and a 5 in the tens place __53__.

17. Give the value of the 5 in each of the following numerals:

    354 __50__
    5,004 __5,000__
    25 __5__
    58,274 __50,000__
    2,577,033 __500,000__

18. How many tens are there in 100? __10__

19. How many hundreds are there in 1,000? __10__

20. How many tens are there in 1,000? __100__

EXPANDED NOTATION

Another way of expressing the place value of a digit is by using the numerical representation of the place value.

The numeral 64

> represents a number where the digit 6 has a value of 6 tens and the digit 4 has a value of 4 ones.

The numeral 64 may be written as

$$6(10) + 4(1) = 60 + 4 = 64$$

You read it as

6 tens + 4 ones = 64

This form of notation is called EXPANDED NOTATION.

The numeral 2,563, written in expanded notation, is

$$2(1000) + 5(100) + 6(10) + 3(1) =$$

$$2{,}000 + 500 + 60 + 3 = 2{,}563$$

The decimal number represented by the expanded notation of

$6(100) + 5(10) + 9(1)$ is

$$600 + 50 + 9 = 659$$

The decimal number represented by the expanded notation of

$7(1000) + 0(100) + 0(10) + 5(1)$ is

$$7{,}000 + 0 + 0 + 5 = 7{,}005$$

HOW TO READ LARGE NUMBERS

To help read and understand numbers of four places or more, we mark off each group of three (3) digits with a COMMA starting from the right.

The last group on the left may have one or two digits instead of three.

For example

5275	becomes	5,275
61347	becomes	61,347
128691	becomes	128,691
7349562	becomes	7,349,562

The first group on the right contains the ONES, TENS and HUNDREDS.
It is called the ONES group.

The second group contains the THOUSANDS, TEN THOUSANDS and the HUNDRED THOUSANDS.
It is called the THOUSANDS group.

The third group contains the MILLIONS, TEN MILLIONS and the HUNDRED MILLIONS.
It is called the MILLIONS group.

The reading of a numeral consists of
1) reading each group as a number
2) followed by the name of the group.

The numeral 27,635 is read as
twenty-seven THOUSAND, six hundred thirty-five.

The "ty" in thirty means "tens."

The numeral 733,682 is read as
seven hundred thirty-three THOUSAND,
six hundred eighty-two.

The numeral 5,227,305 is read as
five MILLION, two hundred twenty-seven
THOUSAND, three hundred five.

Convert the following decimal notations to expanded notation:

1. 45 ____________

2. 178 ____________

3. 3,467 ____________

4. 15,709 ____________

5. 783,552 ____________

Convert from expanded notation to decimal notation:

1. 5 (10) + 4 (1) ____________

2. 3 (100) + 0 (10) + 6 (1) ____________

3. 5 (1000) + 2 (100) + 7 (10) + 0 (1) ____________

Group the digits:

1. 56903 ____________

2. 788304 ____________

3. 1667893 ____________

4. 56637721 ____________

5. 922508347 ____________

6. 1005004006 ____________

Read the following:

1. 5,000 ____________

2. 55,000 ____________

3. 555,000 ____________

4. 5,100 ____________

5. 55,100 ____________

6. 555,100 ____________

7. 5,010 ____________

8. 55,010 ____________

9. 555,010 ____________

10. 5,110 ____________

11. 55,110 ____________

12. 555,110 ____________

13. 5,555,110 ____________

14. 3,636,422,001 ____________

Express the following as numerals:

1. Five thousand thirty-five ____________

2. Eighty-seven thousand one hundred eight ____________

3. Two hundred five thousand sixty-nine ____________

4. Six million six thousand six ____________

5. Five hundred thirty-two million six hundred fifty-five thousand thirty-five ____________

ADDITION–BASIC FACTS

To increase your accuracy and speed.

TRAINING AID

Count aloud the first couple of times; then cover up the numbers and recite aloud Repeat the process again and again, each time increasing your speed.

2, 4, 6, 8, 10, 12, 14, 16, 18, 20

Solve mentally: Add

2 and 2 =	14 and 2 =	9 and 2 =
4 and 2 =	16 and 2 =	11 and 2 =
6 and 2 =	18 and 2 =	13 and 2 =
8 and 2 =	3 and 2 =	15 and 2 =
10 and 2 =	5 and 2 =	17 and 2 =
12 and 2 =	7 and 2 =	19 and 2 =

8	10	5	15	3	13	6	16	7	17	9	19	4	14
2	2	2	2	2	2	2	2	2	2	2	2	2	2

2	2	2	2	2	2	2	2	2	2	2	2	2	2
0	10	1	11	7	20	3	13	9	12	31	21	8	22

Find the missing number:

2 and __ = 6	2 and __ = 4	2 and __ = 8
2 and __ = 3	2 and __ = 9	2 and __ = 2
2 and __ = 7	2 and __ = 5	2 and __ = 10

What number plus 2 will equal 13? ________
What number plus 2 will equal 19? ________
What number plus 2 will equal 15? ________
What number plus 2 will equal 20? ________
What number plus 2 will equal 18? ________

What number plus 2 will equal 12? ________
What number plus 2 will equal 14? ________
What number plus 2 will equal 17? ________
What number plus 2 will equal 16? ________
What number plus 2 will equal 11? ________

ADDITION—BASIC FACTS

TRAINING AID

Count aloud the first couple of times; then cover up the numbers and recite aloud. Repeat the process again and again, each time increasing your speed.

3, 6, 9, 12, 15, 18, 21, 24, 27, 30

Solve mentally: Add

2 and 3 =	14 and 3 =	9 and 3 =
4 and 3 =	16 and 3 =	11 and 3 =
6 and 3 =	18 and 3 =	13 and 3 =
8 and 3 =	3 and 3 =	15 and 3 =
10 and 3 =	5 and 3 =	17 and 3 =
12 and 3 =	7 and 3 =	19 and 3 =

8	18	5	15	3	13	6	16	7	17	9	19	4	14
3	3	3	3	3	3	3	3	3	3	3	3	3	3

3	3	3	3	3	3	3	3	3	3	3	3	3	3
0	10	7	11	6	20	3	23	8	12	9	21	2	22

Find the missing number:

3 and __ = 13	3 and __ = 5	3 and __ = 7
3 and __ = 10	3 and __ = 9	3 and __ = 3
3 and __ = 8	3 and __ = 4	3 and __ = 6

What number plus 3 will equal 14? ________
What number plus 3 will equal 11? ________
What number plus 3 will equal 16? ________
What number plus 3 will equal 17? ________
What number plus 3 will equal 12? ________

What number plus 3 will equal 18? ________
What number plus 3 will equal 15? ________
What number plus 3 will equal 19? ________
What number plus 3 will equal 20? ________
What number plus 3 will equal 13? ________

ADDITION–BASIC FACTS

TRAINING AID

Count aloud the first couple of times; then cover up the numbers and recite aloud. Repeat the process again and again, each time increasing your speed.

4, 8, 12, 16, 20 24, 28, 32, 36, 40

Solve mentally: Add

2 and 4 =	14 and 4 =	9 and 4 =
4 and 4 =	16 and 4 =	11 and 4 =
6 and 4 =	18 and 4 =	13 and 4 =
8 and 4 =	3 and 4 =	15 and 4 =
10 and 4 =	5 and 4 =	17 and 4 =
12 and 4 =	7 and 4 =	19 and 4 =

8	18	5	15	3	13	6	16	7	17	9	19	4	14
4	4	4	4	4	4	4	4	4	4	4	4	4	4

4	4	4	4	4	4	4	4	4	4	4	4	4	4
0	10	7	11	9	20	2	3	23	8	12	1	21	22

Find the missing number:

4 and __ = 7	4 and __ = 5	4 and __ = 17
4 and __ = 4	4 and __ = 10	4 and __ = 12
4 and __ = 8	4 and __ = 6	4 and __ = 14

What number plus 4 will equal 21? ________
What number plus 4 will equal 20? ________
What number plus 4 will equal 16? ________
What number plus 4 will equal 19? ________
What number plus 4 will equal 13? ________

What number plus 4 will equal 18? ________
What number plus 4 will equal 17? ________
What number plus 4 will equal 12? ________
What number plus 4 will equal 15? ________
What number plus 4 will equal 14? ________

ADDITION—BASIC FACTS

TRAINING AID

Count aloud the first couple of times; then cover up the numbers and recite aloud. Repeat the process again and again, each time increasing your speed.

5, 10, 15, 20, 25, 30, 35, 40, 45, 50

Solve mentally: Add

2 and 5 =	14 and 5 =	9 and 5 =
4 and 5 =	16 and 5 =	11 and 5 =
6 and 5 =	18 and 5 =	13 and 5 =
8 and 5 =	3 and 5 =	15 and 5 =
10 and 5 =	5 and 5 =	17 and 5 =
12 and 5 =	7 and 5 =	19 and 5 =

8	18	5	15	3	13	6	16	7	17	9	19	4	14
5	5	5	5	5	5	5	5	5	5	5	5	5	5

5	5	5	5	5	5	5	5	5	5	5	5	5	5
0	10	7	11	8	20	3	23	9	12	1	21	2	22

Find the missing number:

5 and __ = 15	5 and __ = 7	5 and __ = 9
5 and __ = 12	5 and __ = 11	5 and __ = 5
5 and __ = 10	5 and __ = 6	5 and __ = 8

What number plus 5 will equal 16? ________
What number plus 5 will equal 13? ________
What number plus 5 will equal 18? ________
What number plus 5 will equal 19? ________
What number plus 5 will equal 14? ________

What number plus 5 will equal 20? ________
What number plus 5 will equal 17? ________
What number plus 5 will equal 21? ________
What number plus 5 will equal 24? ________
What number plus 5 will equal 15? ________

ADDITION–BASIC FACTS

TRAINING AID

Count aloud the first couple of times; then cover up the numbers and recite aloud. Repeat the process again and again, each time increasing your speed.

6, 12, 18, 24, 30, 36, 42, 48, 54, 60

Solve mentally: Add

2 and 6 =	14 and 6 =	9 and 6 =
4 and 6 =	16 and 6 =	11 and 6 =
6 and 6 =	18 and 6 =	13 and 6 =
8 and 6 =	3 and 6 =	15 and 6 =
10 and 6 =	5 and 6 =	17 and 6 =
12 and 6 =	7 and 6 =	19 and 6 =

8	18	5	15	3	13	6	16	7	17	9	19	4	14
6	6	6	6	6	6	6	6	6	6	6	6	6	6

6	6	6	6	6	6	6	6	6	6	6	6	6	6
0	10	7	11	6	20	3	23	9	12	8	21	2	22

Find the missing number:

6 and ___ = 16	6 and ___ = 8	6 and ___ = 10
6 and ___ = 13	6 and ___ = 12	6 and ___ = 18
6 and ___ = 11	6 and ___ = 7	6 and ___ = 9

What number plus 6 will equal 17? ________
What number plus 6 will equal 14? ________
What number plus 6 will equal 19? ________
What number plus 6 will equal 20? ________
What number plus 6 will equal 15? ________

What number plus 6 will equal 21? ________
What number plus 6 will equal 18? ________
What number plus 6 will equal 22? ________
What number plus 6 will equal 23? ________
What number plus 6 will equal 26? ________

ADDITION–BASIC FACTS

TRAINING AID

Count aloud the first couple of times; then cover up the numbers and recite aloud. Repeat the process again and again, each time increasing your speed.

7, 14, 21, 28, 35, 42, 49, 56, 63, 70

Solve mentally: Add

2 and 7 =	14 and 7 =	9 and 7 =
4 and 7 =	16 and 7 =	11 and 7 =
6 and 7 =	18 and 7 =	13 and 7 =
8 and 7 =	3 and 7 =	15 and 7 =
10 and 7 =	5 and 7 =	17 and 7 =
12 and 7 =	7 and 7 =	19 and 7 =

8	18	5	15	3	13	6	16	7	17	9	19	4	14
7	7	7	7	7	7	7	7	7	7	7	7	7	7

7	7	7	7	7	7	7	7	7	7	7	7	7	7
0	10	7	11	8	20	3	23	9	12	1	21	2	22

Find the missing number:

7 and ___ = 10	7 and ___ = 8	7 and ___ = 12
7 and ___ = 7	7 and ___ = 13	7 and ___ = 14
7 and ___ = 11	7 and ___ = 9	7 and ___ = 17

What number plus 7 will equal 24? ________
What number plus 7 will equal 23? ________
What number plus 7 will equal 19? ________
What number plus 7 will equal 22? ________
What number plus 7 will equal 16? ________

What number plus 7 will equal 21? ________
What number plus 7 will equal 20? ________
What number plus 7 will equal 15? ________
What number plus 7 will equal 18? ________
What number plus 7 will equal 27? ________

ADDITION–BASIC FACTS

TRAINING AID

Count aloud the first couple of times; then cover up the numbers and recite aloud. Repeat the process again and again, each time increasing your speed.

8, 16, 24, 32, 40, 48, 56, 64, 72 80

Solve mentally: Add

2 and 8 =	14 and 8 =	9 and 8 =
4 and 8 =	16 and 8 =	11 and 8 =
6 and 8 =	18 and 8 =	13 and 8 =
8 and 8 =	3 and 8 =	15 and 8 =
10 and 8 =	5 and 8 =	17 and 8 =
12 and 8 =	7 and 8 =	19 and 8 =

8	18	5	15	3	13	6	16	7	17	9	19	4	14
8	8	8	8	8	8	8	8	8	8	8	8	8	8

8	8	8	8	8	8	8	8	8	8	8	8	8	8
0	10	7	11	9	20	3	23	8	12	1	21	2	22

Find the missing number:

8 and ___ = 18	8 and ___ = 10	8 and ___ = 12
8 and ___ = 15	8 and ___ = 14	8 and ___ = 8
8 and ___ = 13	8 and ___ = 9	8 and ___ = 11

What number plus 8 will equal 19? ________
What number plus 8 will equal 16? ________
What number plus 8 will equal 21? ________
What number plus 8 will equal 22? ________
What number plus 8 will equal 17? ________

What number plus 8 will equal 23? ________
What number plus 8 will equal 20? ________
What number plus 8 will equal 24? ________
What number plus 8 will equal 25? ________
What number plus 8 will equal 28? ________

ADDITION–BASIC FACTS

TRAINING AID

Count aloud the first couple of times; then cover up the numbers and recite aloud. Repeat the process again and again, each time increasing your speed.

9, 18, 27, 36, 45, 54, 63, 72, 81, 90

Solve mentally: Add

2 and 9 =	14 and 9 =	9 and 9 =
4 and 9 =	16 and 9 =	11 and 9 =
6 and 9 =	18 and 9 =	13 and 9 =
8 and 9 =	3 and 9 =	15 and 9 =
10 and 9 =	5 and 9 =	17 and 9 =
12 and 9 =	7 and 9 =	19 and 9 =

8	18	5	15	3	13	6	16	7	17	9	19	4	14
9	9	9	9	9	9	9	9	9	9	9	9	9	9

9	9	9	9	9	9	9	9	9	9	9	9	9	9
0	10	7	11	8	20	3	23	6	12	1	21	2	22

Find the missing number:

9 and ___ = 12	9 and ___ = 10	9 and ___ = 14
9 and ___ = 9	9 and ___ = 15	9 and ___ = 16
9 and ___ = 13	9 and ___ = 11	9 and ___ = 19

What number plus 9 will equal 26? ________
What number plus 9 will equal 25? ________
What number plus 9 will equal 21? ________
What number plus 9 will equal 24? ________
What number plus 9 will equal 18? ________

What number plus 9 will equal 23? ________
What number plus 9 will equal 22? ________
What number plus 9 will equal 17? ________
What number plus 9 will equal 19? ________
What number plus 9 will equal 29? ________

ADDITION OF BASIC COMBINATIONS

As soon as man began to arrange his possessions in a definite order, he was COUNTING.

COUNTING leads to the ADDITION of numbers.

Totaling two numbers is a special operation called ADDITION.

Addition permits us to combine two numbers so that they become easier to remember.

THE NUMBER LINE

The number line is a straight line that has been divided into equally spaced units.
Each unit is numbered.
It may be used as a visual method for the addition of numbers.

Example: Add the length of 5 units to a length of 3 units.

1) Proceed as follows:
Draw a line the length of 5 units

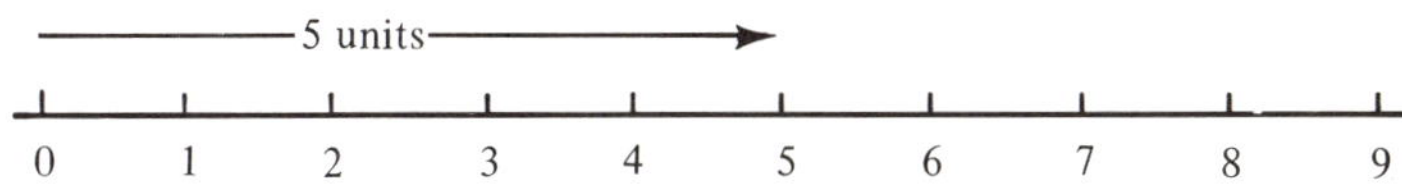

2) From the head of this line draw another line with a length of 3 units

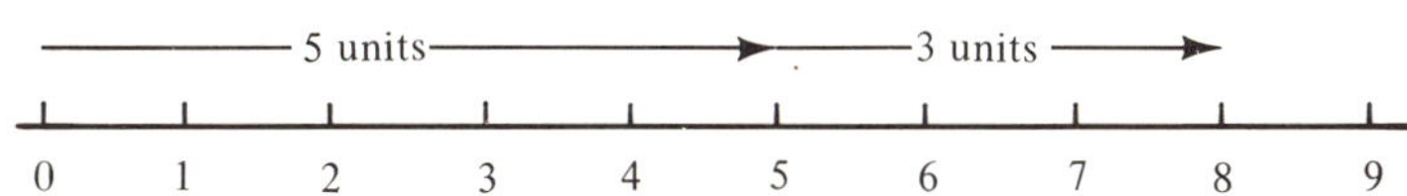

When we add the length of 5 units to a length of 3 units, the overall length becomes 8 units.

In a similar manner we will prove not only that

$$5 + 3 = 8$$

but also that

$$3 + 5 = 8$$

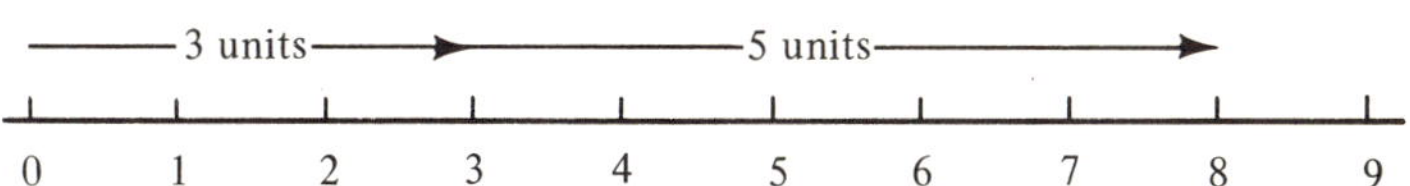

This somewhat insignificant statement that

$$5 + 3 = 3 + 5, \text{ which is } 8,$$

turns out to be one of the most important properties of addition.

This property of addition exists for any two numbers.

This property of addition is called THE COMMUTATIVE PROPERTY.

(To "commute" means "to go back and forth.")

What it boils down to is this:

> You may add a pair of numbers in any order–
>
> downward
> upward
> from left to right
> from right to left
>
> without affecting the sum.

Use the NUMBER LINE to find the following sums:

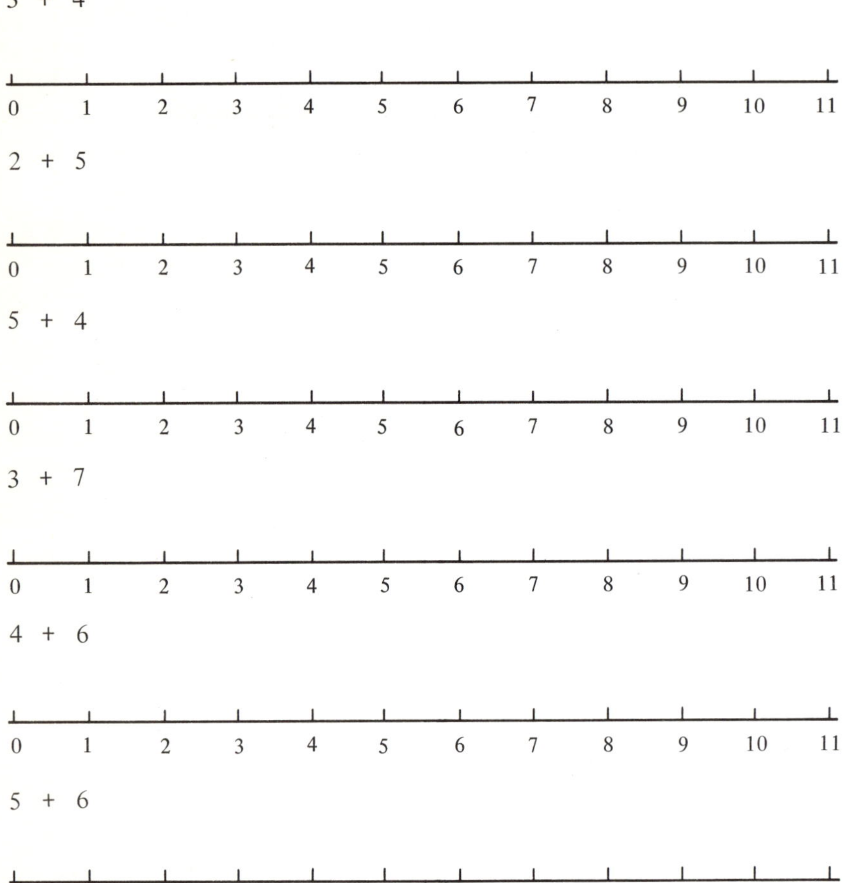

This same procedure may be extended for more than 2 numbers.

NOMENCLATURE

If we want to add five and three we write

$$5 + 3 =$$

The symbol + is called the ADDITION SIGN or the PLUS SIGN.

The numerals 5 and 3 are called the ADDENDS.

The answer to the problem is called the SUM. Some people will call it the TOTAL.

ADDITIVE IDENTITY

If you had \$10 and added nothing more to it, you would still have the \$10.

When zero (0) was added to the \$10 it did not change the value of the \$10.

Zero added to any number preserves the identity of the number.

$$0 + 5 = 5$$

$$7 + 0 = 7$$

The numeral 0 is called the ADDITIVE IDENTITY.

Combine the above property of addition (additive identity) with the commutative property, which states, you may add two numbers in any order without affecting the sum,

and you can reduce the memorization of 90 basic addition facts down to 45.

BASIC ADDITION FACTS

The arrows show a few examples of the Commutative Property.

0 1 1	1 1 2	2 1 3	3 1 4	4 1 5	5 1 6	6 1 7	7 1 8	8 1 9	9 1 10
0 2 2	1 2 3	2 2 4	3 2 5	4 2 6	5 2 7	6 2 8	7 2 9	8 2 10	9 2 11
0 3 3	1 3 4	2 3 5	3 3 6	4 3 7	5 3 8	6 3 9	7 3 10	8 3 11	9 3 12
0 4 4	1 4 5	2 4 6	3 4 7	4 4 8	5 4 9	6 4 10	7 4 11	8 4 12	9 4 13
0 5 5	1 5 6	2 5 7	3 5 8	4 5 9	5 5 10	6 5 11	7 5 12	8 5 13	9 5 14
0 6 6	1 6 7	2 6 8	3 6 9	4 6 10	5 6 11	6 6 12	7 6 13	8 6 14	9 6 15
0 7 7	1 7 8	2 7 9	3 7 10	4 7 11	5 7 12	6 7 13	7 7 14	8 7 15	9 7 16
0 8 8	1 8 9	2 8 10	3 8 11	4 8 12	5 8 13	6 8 14	7 8 15	8 8 16	9 8 17
0 9 9	1 9 10	2 9 11	3 9 12	4 9 13	5 9 14	6 9 15	7 9 16	8 9 17	9 9 18

TIME LIMIT—3 Minutes

ADDITION

As you add these basic combinations, note the combinations that cause you to hesitate. The time limit is there to prevent you from finding the sums by counting with your fingers or any other immature method that will indicate a lack of mastery of the basic facts.

2	5	2	6	2	3	2	7	2	9
4	2	8	2	5	3	0	2	6	2
3	6	3	7	3	4	3	9	3	5
8	3	4	3	0	3	7	3	6	3
4	7	4	9	4	3	4	6	4	7
0	4	6	4	9	4	8	4	7	4
5	3	5	7	5	9	5	6	5	5
2	5	8	5	4	5	0	5	7	9
6	5	6	9	6	3	6	5	6	7
9	6	3	6	8	6	9	6	7	6
7	3	7	8	7	5	7	9	7	0
6	7	5	7	4	7	2	7	8	7
8	4	8	7	8	9	8	3	8	2
3	8	6	8	1	8	0	8	7	8
9	4	9	5	9	0	9	6	9	2
7	9	2	9	3	9	5	9	8	9

SCORING

1 to 4 errors = excellent
5 to 8 errors = good

A TWO-PLACE NUMBER TO A ONE-PLACE NUMBER

Many situations arise in our daily lives that
call for this form of addition;
for example, column addition or
"carrying" when multiplying.

$$\begin{array}{r} 7 \\ 5 \\ \underline{6} \end{array}$$

If we start the addition from the bottom, it would go like this:

6 and 5 is 11
11 and 7 is 18

If we start the addition from the top, it would go like this:

7 and 5 is 12
12 and 6 is 18

As you can see, adding a column of numbers is not just a series of basic facts.

PARTIAL SUMS ARE UNSEEN NUMBERS

In the problem 7 + 5 + 6 = 18

we become involved with numbers like:

12 and 11

These numbers are called PARTIAL SUMS.

They represent a part of the FINAL SUM.

They are not visible, but they are there.

We call them UNSEEN NUMBERS.

Yes, it is true, column addition calls for a skill in adding an INVISIBLE NUMBER to a VISIBLE NUMBER.

It becomes apparent that an integral part of addition is concentration.

You can't wish for it–it must be developed.

To get the ball rolling let's try this.

PRACTICE SUGGESTION

Take a string of numbers such as:

19, 13, 17, 12, 15, 11, 16, 14, 18

Then select a number from 1 to 9–
for example, the number 4–and add it to each of the numbers above to get the sums of

23, 17, 21, 16, 19, 15, 20, 18, 22

Select another number such as 7 and repeat the procedure to get

26, 20, 24, 19, 22, 18, 23, 21, 25

Take another number, such as 5, and repeat;
such as 9, and repeat;
such as 6, and repeat; and so on.

Only practice and more practice will improve your skill in this area–wishful thinking will not do it.

The purpose behind this exercise is to provide you with problems that will require concentration. You may even call this MENTAL ADDITION. We are assuming this is one of your weak areas—and if you're like many of the others with the same problem, you're not going to do anything about it unless someone stands over you with a baseball bat and forces you to practice. Don't look now; we're there in spirit.

ADD 6 to each of the following numbers:

11, 17, 12, 18, 15, 13, 19, 14, 16

— — — — — — — — —

ADD 8 to each of the numbers above to get:

— — — — — — — — —

ADD 5 to each of the numbers above to get:

— — — — — — — — —

ADD 7 to each of the numbers above to get:

— — — — — — — — —

ADD 9 to each of the numbers above to get:

— — — — — — — — —

ADD 7 to each of the following numbers:

26, 24, 29, 23, 25, 28, 22, 27, 21

— — — — — — — — —

ADD 5 to each of the numbers above to get:

— — — — — — — — —

ADD 8 to each of the numbers above to get:

— — — — — — — — —

Mentally add the following problems:

The sum of 3 and 9 added to 5 = ________

The sum of 5 and 7 added to 7 = ________

The sum of 4 and 8 added to 6 = ________

The sum of 7 and 5 added to 8 = ________

The sum of 6 and 7 added to 9 = ________

The sum of 8 and 5 added to 7 = ________

The sum of 12 and 5 added to 7 = ________

The sum of 15 and 6 added to 4 = ________

The sum of 17 and 7 added to 5 = ________

The sum of 14 and 8 added to 8 = ________

The sum of 19 and 7 added to 6 = ________

The sum of 23 and 6 added to 7 = ________

The sum of 25 and 8 added to 7 = ________

The sum of 28 and 8 added to 8 = ________

The sum of 29 and 9 added to 9 = ________

The sum of 34 and 7 added to 9 = ________

The sum of 47 and 6 added to 8 = ________

The sum of 56 and 8 added to 7 = ________

The sum of 64 and 8 added to 6 = ________

The sum of 73 and 8 added to 8 = ________

A TWO-PLACE NUMBER TO A ONE-PLACE NUMBER

This step is so basic to all addition problems that we are going to devote another page to it.

SIMPLIFIED METHOD–The process is clearly mental.

Add

$$\begin{array}{r} 12 \\ \underline{7} \end{array}$$

Think, 7 + 2 = 9

Think, 10 + 9 = 19

To fully understand what took place in the method above, let's analyze it.

If we expand the number 12, we find it contains

12 = 1 ten and 2 ones = 1 (10) + 2 (1)

The number 7 contains

7 = 0 tens and 7 ones = 0 (10) + 7 (1)

Combine the two and we get

$$\begin{array}{rcccccc} 12 & = & 1\,(10) + 2\,(1) & = & 10 + 2 \\ \underline{7} & = & \underline{0\,(10) + 7\,(1)} & = & \underline{0 + 7} \\ 19 & = & 1\,(10) + 9\,(1) & = & 10 + 9 \end{array}$$

ADDITION FROM THE LEFT

The addition can be done even quicker if you start from the LEFT instead of from the right.

Add

$$\begin{array}{r} 22 \\ \underline{6} \end{array}$$

Think, 20 + 8 = 28

If you were to resort to the use of pencil and paper to solve such simple addition problems, it would prove a complete lack of mastery of the basic facts and the first 100 numbers.

The mental approach is what has to be developed.

A slightly different situation exists in the addition of the following problems

Add

$$\begin{array}{r} 25 \\ \underline{9} \end{array}$$

Think, 9 + 5 = 14

Think, 20 + 14 = 34

Visually the problem consists of a two-place number to be added to a one-place number.

Mentally the problem turned out to be the addition of a two-place number to a two-place number.

Let's try it again.

Add

$$\begin{array}{r} 35 \\ \underline{8} \end{array}$$

Think, 8 + 5 = 13

Think, 30 + 13 = 43

ADDITION FROM THE LEFT

Add

$$\begin{array}{r} 47 \\ \underline{9} \end{array}$$

Think, 40 + 16 = 56

For the following short problems perform the addition mentally:

65
4
Think, 4 + 5 = ____
Think, 60 + 9 = ____

83
5
Think, 5 + 3 = ____
Think, 80 + 8 = ____

35
8
Think, 8 + 5 = ____
Think, 30 + 13 = ____

55
4
Think, 4 + 5 = ____
Think, 50 + 9 = ____

86
7
Think, 7 + 6 = ____
Think, 80 + 13 = ____

75
6
Think, 5 + 6 = ____
Think, 70 + 11 = ____

For the following short problems perform the addition from the LEFT:

54
9
Think, 50 + 13 = ____

35
7
Think, 30 + 12 = ____

67
8
Think, 60 + 15 = ____

48
5
Think, 40 + 13 = ____

85
8
Think, 80 + __ = ____

26
7
Think, 20 + __ = ____

37
8
Think, 30 + __ = ____

49
4
Think, 40 + __ = ____

55
9
Think, __ + __ = ____

63
8
Think, __ + __ = ____

76
6
Think, __ + __ = ____

88
9
Think, __ + __ = ____

TIME LIMIT–4 Minutes

Which method did you find easier?
Here's your chance to prove it.
Repeat until you can give the sums without hesitation.

14	16	7	13	11	2	15	11
4	3	12	6	5	14	4	8
23	34	3	32	24	8	22	33
6	4	26	7	5	31	6	5
46	53	3	51	42	8	46	52
2	6	44	8	7	50	3	6
62	76	7	72	61	2	63	76
4	3	62	6	5	74	4	3
14	16	9	13	16	7	15	17
9	7	12	8	9	14	6	5
23	34	7	35	26	8	28	34
8	9	26	6	7	37	5	9
46	55	8	54	47	6	48	59
7	9	47	9	7	56	7	7
68	77	8	79	66	8	64	78
9	5	65	7	7	77	9	6

SCORING

1 to 6 errors = excellent
7 to 12 errors = good

A COLUMN OF ONE-PLACE NUMBERS

What procedure should you use when adding a problem containing three or more numbers?

For example

```
 7
 6
 8
 _
```

If you start from the bottom, the addition should go like this:

8 and 6 is 14
14 and 7 is 21

If you start from the top, the addition should go like this:

7 and 6 is 13
13 and 8 is 21

The above proves that a column of numbers may be added in any manner without affecting the sum.

This important property of addition permits us to check the accuracy of the sum by adding in the opposite direction.

TROUBLE RECALLING PARTIAL SUMS

If your skill in this area is still on the weak side, why not write the partial sums to the right of the column, like this:

```
7        7            7 ]     13
6 ]     14      or    6 ]
8 ]                   8        8
_       __            _       __
```

```
14 + 7 =                      8 + 13 =
Think,  4 +  7 = 11           Think,  3 +  8 = 11
Think, 10 + 11 = 21           Think, 10 + 11 = 21
```

How about a column of four one-place numbers?

```
3          3 ]     8
5          5 ]
9          9 ]    15
6          6 ]
_          _      __
```

We'd like to have it go like this:

```
 6 + 9 = 15
15 + 5 = 20
20 + 3 = 23
```

However, if we paired off the numbers it would go like this:

```
15 + 8  =
 5 + 8  = 13
10 + 13 = 23
```

Let's try another one.

```
8                 8
9       9 ]
7 ]    13 ]      22
6 ]
_      __        __
```

```
13 + 9 =                      22 + 8 =
Think,  3 +  9 = 12           Think,  2 +  8 = 10
Think, 10 + 12 = 22           Think, 20 + 10 = 30
```

Looks rather drawn out, doesn't it?

Of course, we'd like to have it added in the following manner:

Think, 6, 13, 22, 30

and it's going to take practice to get it that way.

PRACTICE SUGGESTION

Every parked and moving vehicle has a set of license plates loaded with numbers—get in the habit of looking for them and adding them.

For example YA-8596

Think, 8, 13, 22, 28

Add the following problems first, from the bottom up and then from the top down:

7	6
5	7
8	8
3	9

Think, 3, ___, 16, ___ ·Think, 9, 17, ___, ___

Think, 7, ___, 20, ___ Think, 6, 13, ___, ___

6	8
8	6
4	6
5	8

Think, 5, ___, 17, ___ Think, 8, 14, ___, ___

Think, 6, ___, 18, ___ Think, 8, 14, ___, ___

8	7
5	7
9	5
6	6

Think, 6, ___, 20, ___ Think, 6, 11, ___, ___

Think, 8, ___, 22, ___ Think, 7, 14, ___, ___

5	9
7	5
4	5
9	8

Think, 9, ___, 20, ___ Think, 8, 13, ___, ___

Think, 5, ___, 16, ___ Think, 9, 14, ___, ___

TIME LIMIT–4 Minutes

Add upward and then check by adding downward:

2	3	4	5	4	3	2
5	6	5	2	1	5	4
4	4	3	4	6	2	3
3	2	1	3	3	4	3

6	7	8	9	8	7	6
2	3	4	5	4	3	2
5	5	4	5	2	5	4
4	4	3	4	6	2	3

2	4	3	5	3	4	2
6	7	8	9	8	7	6
6	7	8	9	8	7	6
2	3	4	5	4	3	4

9	8	7	6	7	8	9
6	7	8	9	8	7	6
2	3	4	5	4	3	2
9	8	7	6	7	8	9

6	7	8	9	8	7	6
9	8	7	6	7	8	9
6	8	7	9	7	8	6
9	8	7	6	8	7	6

SCORING

1 to 3 errors = excellent
4 to 7 errors = good

LONG COLUMN OF ONE-PLACE NUMBERS

Practice in this area is used to register progress.

A characteristic of column addition is the presence of only one combination that is visible while all the others remain unseen.

Concentration is the key to success.

Once you start, don't let anything disturb you.

When you become proficient, the roof could fall in and it would not distract you.

Remember—adding in the opposite direction is a useful check on the accuracy of your work.

PRACTICE PROBLEMS

8
9
6

Think, 6 and 9 is 15
15 and 8 is 23

Check

Think, 8 and 9 is 17
17 and 6 is 23

7
5
9
6

Think, 6 and 9 is 15
15 and 5 is 20
20 and 7 is 27

Check

Think, 7 and 5 is 12
12 and 9 is 21
21 and 6 is 27

8
7
7
6
9

Think, 9 and 6 is 15
15 and 7 is 22
22 and 7 is 29
29 and 8 is 37

Check

Think, 8 and 7 is 15
15 and 7 is 22
22 and 6 is 28
28 and 9 is 37

9
6
8
5
7
6

Think, 6 and 7 is 13
13 and 5 is 18
18 and 8 is 26
26 and 6 is 32
32 and 9 is 41

Check

Think, 9 and 6 is 15
15 and 8 is 23
23 and 5 is 28
28 and 7 is 35
35 and 6 is 41

5
7
9
8
7
6
9
8

Think, 8 and 9 is 17
17 and 6 is 23
23 and 7 is 30
30 and 8 is 38
38 and 9 is 47
47 and 7 is 54
54 and 5 is 59

Check

Think, 5 and 7 is 12
12 and 9 is 21
21 and 8 is 29
29 and 7 is 36
36 and 6 is 42
42 and 9 is 51
51 and 8 is 59

Perform the addition in the prescribed manner:

4
9
7
6

6 and 7 is ___ and 9 is ___ and 4 is ___
4 and 9 is ___ and 7 is ___ and 6 is ___

8
3
8
5

5 and 8 is ___ and 3 is ___ and 8 is ___
8 and 3 is ___ and 8 is ___ and 5 is ___

9
5
9
7

7 and 9 is ___ and 5 is ___ and 9 is ___
9 and 5 is ___ and 9 is ___ and 7 is ___

6
8
5
9

9 and 5 is ___ and 8 is ___ and 6 is ___
6 and 8 is ___ and 5 is ___ and 9 is ___

7
8
8
4
9
6

6 and 9 is ___ and 4 is ___ and 8 is ___
and 8 is ___ and 7 is ___
7 and 8 is ___ and 8 is ___ and 4 is ___
and 9 is ___ and 6 is ___

TIME LIMIT–5 Minutes

Add upward; check by adding downward:

5	2	6	4	7	3
3	6	2	3	3	8
6	4	5	5	3	5
2	3	4	6	4	2
4	6	3	2	6	4
5	1	4	3	2	2

5	8	4	6	7	9
8	2	7	8	3	4
8	7	8	3	5	6
2	8	5	9	8	3
4	5	6	4	2	8
3	6	3	7	6	4

9	4	6	8	7	9
6	8	5	6	5	3
7	7	9	3	9	8
4	9	6	7	8	8
3	6	7	7	4	6
7	3	5	6	8	7

9	8	8	6	7	8
8	7	9	8	9	9
6	7	7	9	9	6
7	9	8	7	8	6
8	8	8	7	8	9
9	6	9	8	7	9

SCORING

1 to 3 errors = excellent
4 to 5 errors = good

A TWO-PLACE NUMBER TO A TWO-PLACE NUMBER

Each two-place number may be regarded as representing a sum:

24	=	2 (tens)	+	4 (ones)	=	20	+	4
15	=	1 (ten)	+	5 (ones)	=	10	+	5
39	=	3 (tens)	+	9 (ones)	=	30	+	9

Such examples should be done MENTALLY.
The addition should begin at the LEFT.
The process involves only TWO STEPS.

Add 24 to 10 to get 34.
Then add the 5 to get 39.

```
2    4
  ↙
1 → 5
```

THINK, two and only two numbers—"34, 39"

Add 35
   27

Add 35 and 20 to get 55.
Then add the 7 to get 62.

```
3    5
  ↙
2 → 7
```

THINK, two and only two numbers—"55, 62"

THE CARRY METHOD

Many students have learned to begin their addition from the RIGHT and then "carry" to the TENS COLUMN, like this:

```
44
28
```

8 + 4 = 12   12 = 1 (10) + 2 (1)
Record 2 in the ones column.
Carry 1 to the tens column.

```
 1
44
28
 2
```

2 + 4 + carried 1 = 7
Record 7 in the tens column.

```
 1
44
28
72
```

THE COLUMN METHOD

And there are many students to whom carrying is a nuisance, and the mental approach requires too much concentration.
So they add in the following manner:

```
56
19
15   =   (9 + 6)
60   =   (50 + 10)
75
```

The COLUMN method has the additional advantage of permitting the addition to begin from the LEFT.

```
56
19
60   =   (50 + 10)
15   =   (9 + 6)
75
```

Why is adding from the left considered important?

In the left-hand column we find the figures with the highest value.
If, in the future, there is a need for an ESTIMATE, finding the sum of that column may be the only adding one needs to do.

Begin each example at the left and add downward. The work should be done mentally. Think two and only two numbers.

23
25 Think 43, ____

45
34 Think 75, ____

54
33 Think ____ , 87

62
27 Think ____ , 89

Add, using the column method, first from the left, and then from the right.

52
35

50 + 30 = ____
2 + 5 = ____
80 + 7 = ____

67
17

7 + 7 = ____
60 + 10 = ____
14 + 70 = ____

Add, using the carry method.

74
19

9 + 4 = ____
Write ___, carry ___
1 + 7 + 1 = ____
Sum = ____

55
37

7 + 5 = ____
Write ___, carry ___
3 + 5 + 1 = ____
Sum = ____

TIME LIMIT—6 Minutes

Try each of the four suggested methods. The one that gets you there the quickest with the mostest is the right one for you.

31	46	13	70	17	67
52	14	54	28	81	22
35	25	42	12	61	26
12	60	17	44	38	50
66	25	59	37	29	24
15	47	39	46	25	57
53	19	29	38	19	17
38	78	14	47	11	68
74	25	56	47	84	28
38	97	58	68	47	76
65	77	33	84	74	19
88	68	78	79	28	88
42	11	15	36	69	48
30	27	57	57	58	84
12	65	64	55	93	96
35	33	19	18	19	17

SCORING

1 to 4 errors = excellent
5 to 8 errors = good

COLUMN OF TWO-PLACE NUMBERS

Skill in this area may serve as a basis for the addition of a column of large numbers. Large problems tend to strain the student's concentration. If that is so, the strain can be relieved by dividing the problem into many parts, summing each part separately and then combining the partial totals.

For example, the following problem may be simplified by dividing it into columns of two-place numbers.

4,318	43	18
2,652	26	52
3,341	33	41
1,857	18	57

THE CRISS-CROSS METHOD

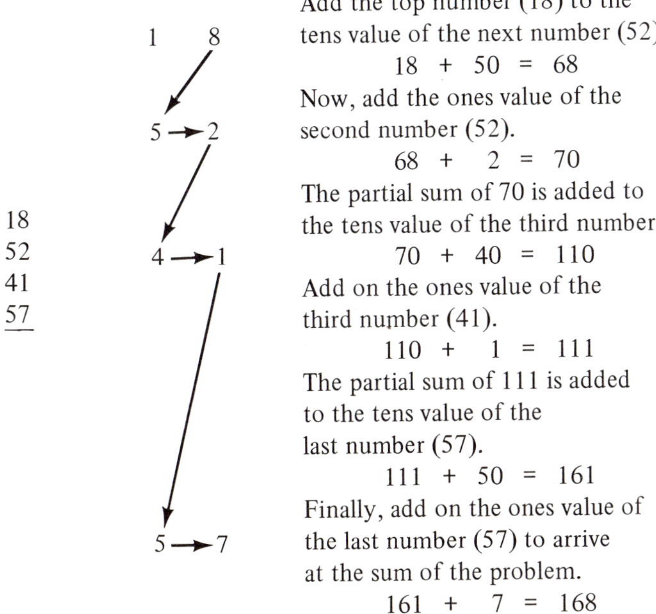

Add the top number (18) to the tens value of the next number (52).

18 + 50 = 68

Now, add the ones value of the second number (52).

68 + 2 = 70

The partial sum of 70 is added to the tens value of the third number.

70 + 40 = 110

Add on the ones value of the third number (41).

110 + 1 = 111

The partial sum of 111 is added to the tens value of the last number (57).

111 + 50 = 161

Finally, add on the ones value of the last number (57) to arrive at the sum of the problem.

161 + 7 = 168

The thinking process should go like this:

18, 68, 70, 110, 111, 161, 168

THE CARRY METHOD

2
43
26
33
18
120

8 + 3 + 6 + 3 = 20 ones

Write 0 in the ones column; carry 2 to the tens column.

1 + 3 + 2 + 4 + 2 = 12 tens

Write 2 in the tens column and 1 in the hundreds column.

THE COLUMN METHOD

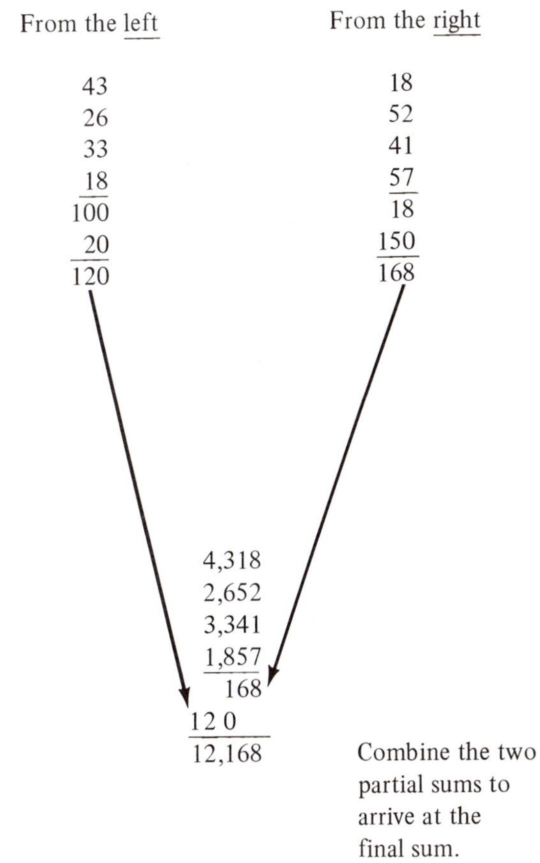

Combine the two partial sums to arrive at the final sum.

PRACTICE PROBLEMS

THE CRISS-CROSS METHOD

28
47
39 Think 68, ___, ___, ___

THE COLUMN METHOD

48
33
67
52

Sum of the ones = ______
Sum of the tens = ______
Sum of the problem = ______

THE CARRY METHOD

55
29
38
66

6 + 8 + 9 + 5 = _______ ones

Write ___ on the ones column
Carry ___ to the tens column

6 + 3 + 2 + 5 + "2" = ___ tens

Sum of the problem = __________

TIME LIMIT—5 Minutes

Record your answers on a piece of paper.
First, use the CRISS-CROSS method.
Then the COLUMN method.
Finally, use the CARRY method.
Compare the time and the errors.

28	36	32	26	33	46
42	32	41	45	53	26
31	44	39	64	27	24
35	24	32	35	44	28
45	46	48	55	36	46
47	37	45	58	45	44
13	38	49	31	31	38
10	49	11	46	38	42
49	50	54	49	56	29
18	22	19	24	36	28
23	45	38	43	11	42
57	36	47	46	43	18
43	35	34	17	48	43
25	32	16	21	27	33
49	48	64	46	35	56
22	56	38	65	25	28
61	31	20	18	65	53
36	29	42	44	56	62
28	28	19	37	54	27
37	32	37	57	65	45
32	53	53	19	37	56

SCORING

1 to 3 errors = excellent
4 to 6 errors = good

COLUMN OF LARGE NUMBERS

The work that follows is as easy as adding a column of one-place or two-place numbers. The procedure(s) will be the same.

THE CARRY METHOD

```
                                          2
                                        3,792
                                        5,268
The sum of the ones column      = 25    6,419
Write 5 in the ones column;             2,856
CARRY 2 to the tens column.             -----
                                            5
```

```
                                         2 2
                                        3,792
                                        5,268
The sum of the tens column      = 23    6,419
Write 3 in the tens column;             2,856
CARRY 2 to the tens column.             -----
                                           35
```

```
                                        2  2 2
                                        3,792
                                        5,268
Sum of the hundreds column      = 23    6,419
Write 3 in the hundreds column;         2,856
CARRY 2 to the thousands column.        -----
                                          335
```

```
                                        2  2 2
                                        3,792
                                        5,268
Sum of the thousands column     = 18    6,419
Write 8 in the thousands column;        2,856
write 1 in the ten thousands column    18,335
```

THE COLUMN METHOD

From the right

```
                                        3,792
                                        5,268
                                        6,419
                                        2,856
                                        -----
6 + 9 + 8 + 2 = 25 ones                    25
5 + 1 + 6 + 9 = 21 tens                   21
8 + 4 + 2 + 7 = 21 hundreds              2 1
2 + 6 + 5 + 3 = 16 thousands            16
                The sum      =         ------
                                       18,335
```

From the left

```
                                        3,792
                                        5,268
                                        6,419
                                        2,856
                                        -----
3 + 5 + 6 + 2 = 16 thousands            16
7 + 2 + 4 + 8 = 21 hundreds              2 1
9 + 6 + 1 + 5 = 21 tens                   21
2 + 8 + 9 + 6 = 25 ones                    25
                The sum      =         ------
                                       18,335
```

Mentally split the problem into two columns of two-place numbers.

```
                                        3,792
Begin the addition with the             5,268
left-hand column of two-place           6,419
numbers and THINK                       2,856
                                        -----
                                       18 1
```

37, 87, 89, 149, 153, 173, 181

(By the way, this addition alone could serve as an ESTIMATE for the problem = 18,100.)

```
                                        3,792
                                        5,268
And now, the right-hand column          6,419
of two-place numbers and THINK          2,856
                                        -----
                                       18 1
                                          235
                                       ------
                                       18,335
```

92, 152, 160, 170, 179, 229, 235

PRACTICE PROBLEMS

CRISS-CROSS METHOD

Mentally split the problem into two columns of two-place numbers.
The choice of starting from the right or left is yours.

	4,213	3,773	2,812
	2,556	2,428	3,141
	1,437	1,892	4,339
	4,159	3,816	2,652
Partial sum =	____	____	____
Partial sum =	____	____	____
Total sum =			

COLUMN METHOD

	3,166	4,328	5,844
	6,646	3,436	4,521
	1,297	6,326	3,747
	2,139	5,162	2,435
Ones =	____	____	____
Tens =	____	____	____
Hundreds =	____	____	____
Thousands =	____	____	____
Sum =			

CARRY METHOD

	3,763	7,854	9,875
	6,843	6,951	2,955
	6,538	8,912	6,670
	2,895	6,714	1,487
Sum =			

TIME LIMIT—10 Minutes

Use the method you prefer.

479	444	179	525	687
625	756	350	649	423
393	551	721	387	314
895	850	498	766	141

565	583	556	688	852
848	347	319	742	658
689	781	752	989	476
288	194	737	346	197

5,824	4,985	8,564	9,325	8,978
7,545	4,567	3,879	3,874	5,776
3,851	7,639	4,279	6,456	2,296
4,668	7,979	5,568	4,539	3,120

5,765	7,585	8,379	3,994	9,311
2,795	3,823	6,429	9,436	6,525
8,495	5,092	5,172	8,294	2,018
4,817	8,319	2,352	2,547	2,432

766,324	64,964,663
443,152	83,355,854
245,864	18,651,748
67,177	18,656,427

SCORING

1 to 2 errors = excellent
3 to 4 errors = good

ESTIMATING SUMS

To estimate is to look at a problem and approximate the size of the answer.
To approximate means getting close to the correct answer.
Estimating should be done quickly and simply.
Complicated-looking numbers should be simplified.
Simplified numbers are numbers that have been rounded off to the nearest TEN, HUNDRED, and so forth.

ROUNDING OFF TO THE NEAREST TEN

47 should be rounded off to 50
because it is closer to
50 than it is to 40.

44 should be rounded off to 40
because it is closer to
40 than it is to 50.

247 should be rounded off to 250
because it is closer to
250 than it is to 240.

274 should be rounded off to 270
because it is closer to
270 than it is to 280.

ROUNDING OFF TO THE NEAREST HUNDRED

628 should be rounded off to 600
651 should be rounded off to 700

2,392 should be rounded off to 2,400
2,449 should be rounded off to 2,400

ROUNDING OFF TO THE NEAREST THOUSAND

3,456 should be rounded off to 3,000
9,975 should be rounded off to 10,000
12,455 should be rounded off to 12,000
249,544 should be rounded off to 250,000

SAMPLE EXERCISE

34,493 rounded off to the nearest thousand.......... 34,000
34,493 rounded off to the nearest hundred 34,500
34,493 rounded off to the nearest ten 34,490

IS 65 CLOSER TO 70 THAN IT IS TO 60?

RULE: If the second-place digit is
5 or greater than 5
INCREASE the first-place digit by 1

If the second-place digit is
4 or less than 4
DO NOT INCREASE the first-place digit

65 should be rounded off to 70

64 should be rounded off to 60

ESTIMATING SUMS BY ADDING FIRST-PLACE DIGITS

56
34
27
85

Estimated sum =
5 + 3 + 2 + 8 = 18 tens
18 tens = 180

ESTIMATING SUMS BY ROUNDING OFF

56	60
34	30
27	30
85	90
	210 = estimated sum

Find the correct sum:

56
34
27
85
202

CONCLUSION: Estimating sums by rounding off can come very close to the correct sum.

Round off to the nearest TEN:

166 __________	284 __________
236 __________	438 __________
665 __________	745 __________
1,345 __________	2,732 __________

Round off to the nearest HUNDRED:

652 __________	346 __________
648 __________	750 __________
5,240 __________	2,557 __________
5,742 __________	4,747 __________

Round off to the nearest THOUSAND:

3,375 __________	313,313 __________
8,759 __________	858,518 __________
25,400 __________	4,627,738 __________
56,613 __________	8,841,500 __________

2,390 is rounded off to the nearest ____________________ .

2,400 is rounded off to the nearest ____________________ .

2,000 is rounded off to the nearest ____________________ .

67,000 is rounded off to the nearest ____________________ .

67,300 is rounded off to the nearest ____________________ .

67,340 is rounded off to the nearest ____________________ .

445,350 is rounded off to the nearest ____________________ .

445,400 is rounded off to the nearest ____________________ .

445,000 is rounded off to the nearest ____________________ .

450,000 is rounded off to the nearest ____________________ .

500,000 is rounded off to the nearest ____________________ .

	Estimate the sum by	
35	Adding first-place digits	= __________
62	Rounding off	= __________
47	Actual sum	= __________
56		

	Estimate the sum by	
248	Adding first-place digits	= __________
563	Rounding off to the nearest hundred	= __________
379	Actual sum	= __________
832		

	Estimate the sum by	
4,762	Adding first-place digits	= __________
6,392	Rounding off to the nearest thousand	= __________
8,517	Actual sum	= __________
3,642		

	Estimate the sum by	
47,259	Adding first-place digits	= __________
69,144	Rounding off to the nearest ten thousand	= __________
33,837	Rounding off to the nearest thousand	= __________
88,296	Actual sum	= __________

	Estimate the sum by	
897,272	Adding first-place digits	= __________
484,951	Rounding off to the nearest hundred thousand	= __________
502,771	Rounding off to the nearest ten thousand	= __________
341,167	Actual sum	= __________

	Estimate the sum by	
4,828,128	Adding first-place digits	= __________
2,761,481	Rounding off to the nearest million	= __________
7,906,908	Rounding off to the nearest hundred thousand	= __________
8,191,622	Actual sum	= __________

TIME LIMIT—10 Minutes

ACHIEVEMENT TEST 1

9	8	7	9	7	8	7
8	6	9	8	6	4	9
			9	8	6	2
			7	4	9	8

48	69	57	45	64	76	52
17	28	39	73	43	42	36
			54	59	23	44
			28	78	91	75

90	50	40	170	760	504	706
60	30	70	280	950	809	402
80	70	90	350	370	708	209
40	90	90				

1,040	631	6,070	716	903,500	76,026
5,209	764	8,002	482	8,020	95,700
8,009	395	2,103	213	35,608	80,205
7,080		9,051		406,700	
2,800		3,503		780,093	

Add horizontally and vertically:

92 + 89 + 43 + 67 + 38 = ______

44 + 51 + 83 + 68 + 29 = ______

19 + 37 + 47 + 92 + 58 = ______

57 + 29 + 52 + 38 + 35 = ______

16 + 74 + 53 + 69 + 26 = ______

36 + 64 + 97 + 46 + 37 = ______

___ ___ ___ ___ ___ ______

Total ______

The vertical sum must equal the horizontal sum.

Purchases at the supermarket

Steak	$3.59	Lettuce	$ .69	Ice cream	$ .86
Potatoes	.87	Cheese	1.36	Cold cuts	4.95
Onions	.35	Cake	.78	Oil	.77
Bread	.72	Cookies	.49	Vinegar	.39
Milk	1.59	Soda	1.33	Napkins	.28
				Total cost	______

At the hardware store

Hammer	$2.59
Pliers	1.69
Nails	.56
Saw	3.78

At the department store

4 yards of cloth	$9.16
3 yards of lining	2.39
1 pattern	.45
Thread	.39
Buttons	.59

Total spent for the day ______

Expenses for the family trip

Gasoline	$29.37
Oil	2.70
Motel	72.00
Food	76.78
Refreshments	9.75
Tire repairs	6.80
Tolls	5.55
Total expenses	

Monthly installment payments

Home	$198.10
Car	125.75
Color TV	48.87
Washing machine	25.69
Rugs	37.94
Total payments	

RELATED PROBLEMS–ADDITION

PAYROLL

	Monday	Tuesday	Wednesday	Thursday	Friday		Totals per man
Jones	$38.72	$36.87	$41.19	$34.65	$40.89	=	______
Smith	29.57	30.91	26.58	32.38	34.17	=	______
Allen	33.89	29.47	31.68	34.92	30.85	=	______
Brown	26.18	28.49	25.84	27.69	32.39	=	______
Slater	27.55	30.08	26.77	25.38	31.18	=	______
Muller	36.84	34.37	29.59	31.93	35.67	=	______
Stevens	42.16	36.81	37.09	40.87	43.56	=	______
Totals per day	______	______	______	______	______	=	______ Total per week

NOTE:

The vertical sum must equal the horizontal sum.

Attendance and receipts at the Stadium

Monday	42,653	$267,562.94
Tuesday	36,089	197,728.39
Wednesday	44,918	296,994.47
Thursday	33,836	164,827.18
Friday	35,649	176,409.08
Saturday	49,058	243,428.76
Sunday	51,776	279,664.65
Total attendance	______	______ Total receipts

Estimated U.S. population by age group for 1980

Under 5 years	31,040,000
5 to 13	45,215,000
14 to 17	16,005,000
18 to 24	29,612,000
25 to 34	36,998,000
35 to 44	25,376,000
45 to 54	22,147,000
55 to 64	21,032,000
65 to 74	14,457,000
75 and over	8,606,000
All ages	______

Chapter 2

SUBTRACTION

PRETEST–SUBTRACTION

The most common faults in subtraction are listed below:

Heading the list is "errors in basic facts."

The next common fault is difficulty in regrouping (most people call it "borrowing").

Lack of comprehension of proper procedures, lack of concentration.

Failure to estimate the answer–a rough answer to be used as a check against the answer.

And finally, failing to check the accuracy of the answer when subtraction affords the individual a simple, foolproof method–providing his skill in addition is excellent.

Because corrective measures vary from student to student, remedial treatment must proceed on a step-by-step basis at a level that promises success and thereby provides for an enjoyable and profitable learning experience.

Take the pretest.

If you can answer 23 out of 25, you may, if you wish, bypass this chapter–however, we feel it will be worth your while to have a look.

PRETEST–SUBTRACTION

48	60	55	63	256	749	714	802	800
–27	–32	–20	–38	–203	–709	–358	–272	–747

1,633	8,000	8,070	7,009	8,074
–874	–5,080	–3,800	–3,900	–2,295

7,337	6,285	9,000	7,020	7,006
–4,139	–4,369	–5,886	–4,693	–3,058

258,206	91,020	18,036	23,007
–88,638	–90,286	–7,797	–4,009

387,364	1,002,005
–89,965	–998,998

SUBTRACTION—BASIC FACTS

TRAINING AID

Subtraction is the inverse of addition. Memorization of the basic addition tables is an absolute necessity.

REVIEW—Addition table

2, 4, 6, 8, 10, 12, 14, 16, 18, 20

Solve mentally: Subtract

2 from 2 =	2 from 8 =	2 from 14 =
2 from 4 =	2 from 10 =	2 from 16 =
2 from 6 =	2 from 12 =	2 from 18 =
2 from 3 =	2 from 9 =	2 from 15 =
2 from 5 =	2 from 11 =	2 from 17 =
2 from 7 =	2 from 13 =	2 from 19 =

8	18	5	15	3	13	6	16	7	17	9	19
−2	−2	−2	−2	−2	−2	−2	−2	−2	−2	−2	−2

4	14	10	20	21	12	2	22	23	25	33	11
−2	−2	−2	−2	−2	−2	−2	−2	−2	−2	−2	−2

Find the missing number:

___ minus 2 = 10	___ minus 2 = 5	___ minus 2 = 8
___ minus 2 = 3	___ minus 2 = 7	___ minus 2 = 9
___ minus 2 = 6	___ minus 2 = 4	___ minus 2 = 0

2 subtracted from what number = 11? ________
2 subtracted from what number = 13? ________
2 subtracted from what number = 16? ________
2 subtracted from what number = 15? ________
2 subtracted from what number = 18? ________

2 subtracted from what number = 12? ________
2 subtracted from what number = 14? ________
2 subtracted from what number = 16? ________
2 subtracted from what number = 17? ________
2 subtracted from what number = 20? ________

SUBTRACTION–BASIC FACTS

TRAINING AID

Subtraction is the inverse of addition. Memorization of the basic addition tables is an absolute necessity.

REVIEW–Addition table

3, 6, 9, 12, 15, 18, 21, 24, 27, 30

Solve mentally: Subtract

3 from 3 =	3 from 6 =	3 from 9 =
3 from 12 =	3 from 15 =	3 from 18 =
3 from 21 =	3 from 24 =	3 from 27 =
3 from 30 =	3 from 33 =	3 from 36 =
3 from 11 =	3 from 10 =	3 from 13 =
3 from 8 =	3 from 7 =	3 from 5 =

8	18	5	15	6	16	7	17	9	19	4	14	13
-3	-3	-3	-3	-3	-3	-3	-3	-3	-3	-3	-3	-3

26	10	21	11	20	22	12	23	3	33	24	28	25
-3	-3	-3	-3	-3	-3	-3	-3	-3	-3	-3	-3	-3

Find the missing number:

___ minus 3 = 0	___ minus 3 = 10	___ minus 3 = 9
___ minus 3 = 5	___ minus 3 = 8	___ minus 3 = 4
___ minus 3 = 9	___ minus 3 = 6	___ minus 3 = 11

3 subtracted from what number = 18? ________	3 subtracted from what number = 13? ________
3 subtracted from what number = 19? ________	3 subtracted from what number = 17? ________
3 subtracted from what number = 16? ________	3 subtracted from what number = 14? ________
3 subtracted from what number = 20? ________	3 subtracted from what number = 12? ________
3 subtracted from what number = 15? ________	3 subtracted from what number = 9? ________

SUBTRACTION–BASIC FACTS

TRAINING AID

Subtraction is the inverse of addition. Memorization of the basic addition tables is an absolute necessity.

REVIEW–Addition table

4, 8, 12, 16, 20, 24, 28, 32, 36, 40

Solve mentally: Subtract

4 from 4 =	4 from 24 =	4 from 8 =
4 from 28 =	4 from 12 =	4 from 32 =
4 from 16 =	4 from 36 =	4 from 20 =
4 from 40 =	4 from 13 =	4 from 23 =
4 from 9 =	4 from 17 =	4 from 33 =
4 from 21 =	4 from 29 =	4 from 31 =

8	18	5	15	4	14	6	16	7	17	9	19
–4	–4	–4	–4	–4	–4	–4	–4	–4	–4	–4	–4

12	10	21	20	22	25	23	13	30	33	24	11
–4	–4	–4	–4	–4	–4	–4	–4	–4	–4	–4	–4

Find the missing number:

__ minus 4 = 12	__ minus 4 = 6	__ minus 4 = 10
__ minus 4 = 4	__ minus 4 = 5	__ minus 4 = 0
__ minus 4 = 7	__ minus 4 = 9	__ minus 4 = 8

4 subtracted from what number = 18? ________	4 subtracted from what number = 13? ________
4 subtracted from what number = 19? ________	4 subtracted from what number = 17? ________
4 subtracted from what number = 16? ________	4 subtracted from what number = 14? ________
4 subtracted from what number = 20? ________	4 subtracted from what number = 12? ________
4 subtracted from what number = 15? ________	4 subtracted from what number = 11? ________

SUBTRACTION—BASIC FACTS

TRAINING AID

Subtraction is the inverse of addition.
Memorization of the basic addition tables is an absolute necessity.

REVIEW—Addition table

5, 10, 15, 20, 25, 30, 35, 40, 45, 50

Solve mentally: Subtract

5 from 5 =	5 from 30 =	5 from 10 =
5 from 35 =	5 from 15 =	5 from 40 =
5 from 20 =	5 from 45 =	5 from 25 =
5 from 50 =	5 from 11 =	5 from 23 =
5 from 17 =	5 from 24 =	5 from 19 =
5 from 27 =	5 from 31 =	5 from 28 =

8 −5	18 −5	5 −5	15 −5	9 −5	19 −5	7 −5	17 −5	6 −5	16 −5	11 −5	12 −5
26 −5	10 −5	21 −5	14 −5	24 −5	13 −5	23 −5	25 −5	20 −5	22 −5	28 −5	30 −5

Find the missing number:

___ minus 5 = 0	___ minus 5 = 6	___ minus 5 = 8
___ minus 5 = 11	___ minus 5 = 9	___ minus 5 = 5
___ minus 5 = 10	___ minus 5 = 7	___ minus 5 = 13

5 subtracted from what number = 13? ________
5 subtracted from what number = 15? ________
5 subtracted from what number = 18? ________
5 subtracted from what number = 14? ________
5 subtracted from what number = 16? ________

5 subtracted from what number = 21? ________
5 subtracted from what number = 17? ________
5 subtracted from what number = 20? ________
5 subtracted from what number = 19? ________
5 subtracted from what number = 22? ________

SUBTRACTION—BASIC FACTS

TRAINING AID

Subtraction is the inverse of addition. Memorization of the basic addition tables is an absolute necessity.

REVIEW—Addition table

6, 12, 18, 24, 30, 36, 42, 48, 54, 60

Solve mentally: Subtract

6 from 6 =	6 from 36 =	6 from 12 =
6 from 42 =	6 from 18 =	6 from 48 =
6 from 24 =	6 from 54 =	6 from 24 =
6 from 60 =	6 from 17 =	6 from 21 =
6 from 27 =	6 from 31 =	6 from 28 =
6 from 18 =	6 from 20 =	6 from 29 =

8	18	6	16	9	19	7	17	14	16	12	14
−6	−6	−6	−6	−6	−6	−6	−6	−6	−6	−6	−6

23	10	21	11	20	22	12	13	30	15	33	24
−6	−6	−6	−6	−6	−6	−6	−6	−6	−6	−6	−6

Find the missing number:

___ minus 6 = 14	___ minus 6 = 8	___ minus 6 = 11
___ minus 6 = 6	___ minus 6 = 10	___ minus 6 = 12
___ minus 6 = 9	___ minus 6 = 7	___ minus 6 = 0

6 subtracted from what number = 20? ________
6 subtracted from what number = 21? ________
6 subtracted from what number = 18? ________
6 subtracted from what number = 22? ________
6 subtracted from what number = 17? ________

6 subtracted from what number = 15? ________
6 subtracted from what number = 19? ________
6 subtracted from what number = 16? ________
6 subtracted from what number = 14? ________
6 subtracted from what number = 23? ________

SUBTRACTION–BASIC FACTS

TRAINING AID

Subtraction is the inverse of addition. Memorization of the basic addition tables is an absolute necessity.

REVIEW–Addition table

7, 14, 21, 28, 35, 42, 49, 56, 63, 70

Solve mentally: Subtract

7 from 7 =	7 from 42 =	7 from 14 =
7 from 49 =	7 from 21 =	7 from 56 =
7 from 28 =	7 from 63 =	7 from 35 =
7 from 70 =	7 from 17 =	7 from 24 =
7 from 36 =	7 from 29 =	7 from 18 =
7 from 43 =	7 from 32 =	7 from 25 =

8	18	7	17	9	19	11	16	12	15	13	14
–7	–7	–7	–7	–7	–7	–7	–7	–7	–7	–7	–7

25	10	21	31	20	30	22	32	23	33	24	34
–7	–7	–7	–7	–7	–7	–7	–7	–7	–7	–7	–7

Find the missing number:

__ minus 7 = 0	__ minus 7 = 8	__ minus 7 = 10
__ minus 7 = 13	__ minus 7 = 11	__ minus 7 = 7
__ minus 7 = 12	__ minus 7 = 9	__ minus 7 = 15

7 subtracted from what number = 15? ________
7 subtracted from what number = 17? ________
7 subtracted from what number = 20? ________
7 subtracted from what number = 16? ________
7 subtracted from what number = 21? ________

7 subtracted from what number = 18? ________
7 subtracted from what number = 23? ________
7 subtracted from what number = 19? ________
7 subtracted from what number = 22? ________
7 subtracted from what number = 25? ________

SUBTRACTION–BASIC FACTS

TRAINING AID

Subtraction is the inverse of addition. Memorization of the basic addition tables is an absolute necessity.

REVIEW–Addition table

8, 16, 24, 32, 40, 48, 56, 64, 72, 80

Solve mentally: Subtract

8 from 8 =	8 from 48 =	8 from 16 =
8 from 56 =	8 from 24 =	8 from 64 =
8 from 32 =	8 from 72 =	8 from 40 =
8 from 80 =	8 from 21 =	8 from 33 =
8 from 19 =	8 from 42 =	8 from 45 =
8 from 20 =	8 from 30 =	8 from 50 =

8	18	9	19	17	11	16	12	15	13	14	24
–8	–8	–8	–8	–8	–8	–8	–8	–8	–8	–8	–8

40	10	21	30	22	25	35	23	33	31	26	32
–8	–8	–8	–8	–8	–8	–8	–8	–8	–8	–8	–8

Find the missing number:

___ minus 8 = 16	___ minus 8 = 10	___ minus 8 = 13
___ minus 8 = 8	___ minus 8 = 12	___ minus 8 = 14
___ minus 8 = 11	___ minus 8 = 9	___ minus 8 = 0

8 subtracted from what number = 22? ________
8 subtracted from what number = 23? ________
8 subtracted from what number = 20? ________
8 subtracted from what number = 24? ________
8 subtracted from what number = 19? ________

8 subtracted from what number = 17? ________
8 subtracted from what number = 21? ________
8 subtracted from what number = 18? ________
8 subtracted from what number = 16? ________
8 subtracted from what number = 25? ________

SUBTRACTION–BASIC FACTS

TRAINING AID

Subtraction is the inverse of addition.
Memorization of the basic addition tables is an absolute necessity.

REVIEW–Addition table

9, 18, 27, 36, 45, 54, 63, 72, 81, 90

Solve mentally: Subtract

9 from 9 =	9 from 54 =	9 from 18 =
9 from 63 =	9 from 27 =	9 from 72 =
9 from 36 =	9 from 81 =	9 from 45 =
9 from 90 =	9 from 25 =	9 from 37 =
9 from 42 =	9 from 19 =	9 from 31 =
9 from 53 =	9 from 38 =	9 from 40 =

9	19	17	11	16	12	15	13	14	29	25	35
−9	−9	−9	−9	−9	−9	−9	−9	−9	−9	−9	−9

18	10	21	26	20	22	28	33	30	24	29	34
−9	−9	−9	−9	−9	−9	−9	−9	−9	−9	−9	−9

Find the missing number:

__ minus 9 = 0	__ minus 9 = 10	__ minus 9 = 9
__ minus 9 = 15	__ minus 9 = 13	__ minus 9 = 12
__ minus 9 = 14	__ minus 9 = 11	__ minus 9 = 17

9 subtracted from what number = 17? ________
9 subtracted from what number = 19? ________
9 subtracted from what number = 22? ________
9 subtracted from what number = 18? ________
9 subtracted from what number = 20? ________

9 subtracted from what number = 25? ________
9 subtracted from what number = 21? ________
9 subtracted from what number = 24? ________
9 subtracted from what number = 23? ________
9 subtracted from what number = 30? ________

WHAT IS SUBTRACTION?

Subtraction is the name of the process that TAKES AWAY one quantity from another.

Example: Take 3 apples away from 5 apples.

Subtraction is the name of the process that finds the DIFFERENCE between two quantities.

Example: The difference between a 10-inch ruler and 12-inch ruler is _____.

Subtraction is the name of the process that finds the REMAINDER after a withdrawal has been made.

Example: If you withdraw $10 from a balance of $25 you will have a remainder of _____.

Subtraction is the name of the process that finds how much larger one quantity is than another—or how much smaller one quantity is than another.

Example: If Mary is 5 feet tall and John is 6 feet tall, how much taller is John than Mary?

Mary is how much smaller than John?

Subtraction is also the name of the process that can find how much more must be added to a smaller quantity to equal a larger quantity.

Example: What number added to 4 will equal 10?

SUBTRACTION IS AN INVERSE OPERATION

Inverse means the direct opposite, the reverse.
Counting forward is called ADDITION.
Counting backward is called SUBTRACTION.
This characteristic will permit you to solve subtraction problems by addition.

SUBTRACTION PERFORMED ON THE NUMBER LINE

Subtract the length of 3 units from a length of 8 units.
Proceed as follows:
1) Draw a line the length of 8 units.

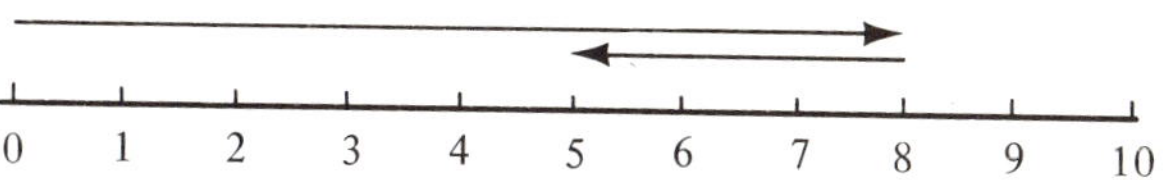

2) From the head of this line draw another line to the LEFT (remember, subtraction is counting backward) 3 units long.

When we subtract 3 units from a length of 8 units the REMAINDER length becomes 5 units.

NOMENCLATURE

If we want to subtract 3 from 8 to get a remainder of 5, we write:

$$\begin{array}{r} 8 \\ -3 \\ \hline 5 \end{array}$$

The symbol – is called the MINUS SIGN.
8 is called the MINUEND.
3 is the called the SUBTRAHEND ("sub" means "under").
5 is called the REMAINDER, or the DIFFERENCE.

MORE FACTS ABOUT SUBTRACTION

0 subtracted from any number does not change the value of the number

$$\begin{array}{r} 8 \\ -0 \\ \hline 8 \end{array}$$

The difference between two equal numbers is 0.

$$\begin{array}{r} 8 \\ -8 \\ \hline 0 \end{array}$$

The accuracy of the remainder can always be checked by adding it to the subtrahend to equal the minuend: $8 - 3 = 5,\ 5 + 3 = 8$

Use the NUMBER LINE to find the difference.

7 − 4 =

0 1 2 3 4 5 6 7 8 9 10 11 12 13 14

9 − 3 =

0 1 2 3 4 5 6 7 8 9 10 11 12 13 14

11 − 6 =

0 1 2 3 4 5 6 7 8 9 10 11 12 13 14

13 − 5 =

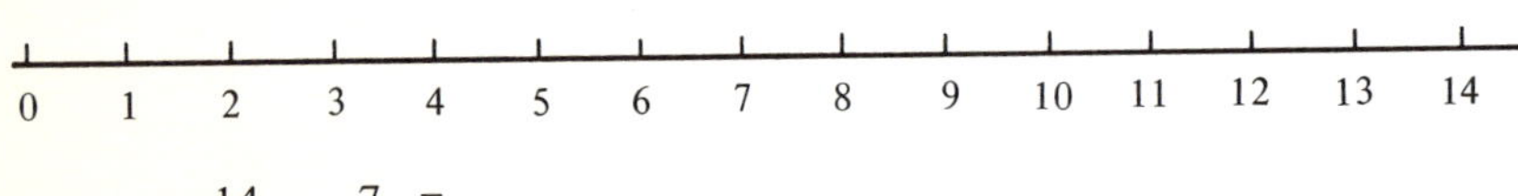

14 − 7 =

0 1 2 3 4 5 6 7 8 9 10 11 12 13 14

10 − 3 =

0 1 2 3 4 5 6 7 8 9 10 11 12 13 14

12 − 5 =

0 1 2 3 4 5 6 7 8 9 10 11 12 13 14

TIME LIMIT—3 Minutes

The responses should not be thought over; they should be instinctive.

11	15	12	6	12	8	9	12	15	16
−6	−9	−4	−2	−9	−1	−0	−6	−7	−8
18	11	6	3	8	15	7	9	14	13
−9	−8	−4	−2	−3	−5	−5	−9	−7	−8
6	9	12	16	10	5	6	16	12	11
−2	−4	−7	−9	−3	−3	−1	−8	−7	−8
6	12	14	17	8	6	8	12	17	18
−0	−5	−7	−9	−4	−0	−5	−4	−7	−8
9	6	11	8	3	10	4	10	11	19
−5	−4	−3	−1	−0	−3	−0	−1	−7	−8
7	13	8	6	4	13	6	11	15	14
−7	−5	−1	−0	−4	−7	−6	−7	−7	−8
8	4	9	14	6	15	5	16	19	10
−3	−0	−5	−8	−3	−6	−2	−8	−7	−8
15	13	17	12	11	10	14	16	18	20
−1	−2	−3	−4	−5	−6	−7	−8	−9	−9
15	13	17	12	11	10	14	16	18	20
−9	−8	−7	−6	−5	−4	−3	−2	−1	−4

SCORING

1 to 4 errors = excellent
5 to 8 errors = good

A ONE-PLACE NUMBER FROM A TWO-PLACE NUMBER

Subtract

$$\begin{array}{r} 18 \\ -6 \\ \hline \end{array}$$

If we analyze the number 18 we find it contains

1 ten + 8 ones = 1 (10) + 8 (1)

The number 6 contains

0 tens + 6 ones = 0 (10) + 6 (1)

Combine the two and we get

$$\begin{array}{rcccccccc} 18 & = & 1(10) & + & 8(1) & = & 10 & + & 8 \\ -6 & = & 0(10) & + & 6(1) & = & 0 & + & 6 \\ \hline 12 & = & 1(10) & + & 2(1) & = & 10 & + & 2 \end{array}$$

THE SHORT FORM (Practiced by most)

$$\begin{array}{r} 18 \\ -6 \\ \hline 12 \end{array}$$

Begin the subtraction with the ones column.

6 from 8 = 2

Now, switch over to the tens column.
(Remember: 0 subtracted from any number does not change the value of the number.)

0 from 1 = 1

ADDITIVE-SUBTRACTIVE METHOD

This method makes short work out of small problems such as:

$$\begin{array}{r} 18 \\ -6 \\ \hline 12 \end{array}$$

You find the difference by asking,
"What number when added to 6 will equal 18?"

Of course, the answer is 12.

The subtraction problem was converted into an addition problem.

CHECKING THE ANSWER

When the sum of the difference and the subtrahend equals the minuend, the answer is correct.

18 – 6 = 12

because 12 + 6 = 18

TIME LIMIT–5 Minutes

SUBTRACTION

First time around, use the SHORT-FORM method.
Note the combinations that cause you to hesitate.
The responses should not be thought over; they should be instinctive.
Second time around, try the ADDITIVE-SUBTRACTIVE method.

SCORING

1 to 4 errors = excellent
5 to 8 errors = good

19	25	36	47	58	69	78	87	96
–2	–2	–2	–2	–2	–2	–2	–2	–2
18	26	37	48	59	68	77	86	95
–3	–3	–3	–3	–3	–3	–3	–3	–3
17	27	38	49	58	67	76	85	94
–4	–4	–4	–4	–4	–4	–4	–4	–4
16	28	39	48	57	66	75	86	95
–5	–5	–5	–5	–5	–5	–5	–5	–5
15	29	38	47	56	67	76	87	96
–6	–6	–6	–6	–6	–6	–6	–6	–6
14	28	37	46	55	64	73	84	95
–10	–10	–10	–10	–10	–10	–10	–10	–10
15	27	36	45	54	63	74	85	96
–3	–3	–3	–3	–3	–3	–3	–3	–3
16	26	35	44	53	64	75	86	97
–2	–2	–2	–2	–2	–2	–2	–2	–2
17	25	34	45	54	65	76	87	98
–4	–4	–4	–4	–4	–4	–4	–4	–4
18	26	35	46	55	66	77	88	99
–5	–5	–5	–5	–5	–5	–5	–5	–5

BORROWING (Regrouping)

When is borrowing necessary?
Whenever within a given place value, the digit in the subtrahend is larger than the digit in the minuend.

For example:

$$\begin{array}{r} 35 \\ \underline{-9} \end{array} = \begin{array}{l} 3(10) + 5(1) \\ \underline{0(10) + 9(1)} \end{array} = \begin{array}{r} 30 + 5 \\ \underline{0 + 9} \end{array}$$

You cannot subtract 9 ones from 5 ones. But if you "borrow" 1 ten from the 3 tens and exchange it for 10 ones–add it to the 5 ones to get 15 ones–then you will be ready to subtract.

The same problem "regrouped":

35 regrouped	= 2 (10) + 15 (1)	=	20 + 15
–9	= 0 (10) + 9 (1)	=	0 + 9
26	= 2 (10) + 6 (1)	=	20 + 6

9 ones from 15 ones = 6 ones

0 tens from 2 tens = 2 tens

2 (10) + 6 (1) = 26 ones

ADDITIVE-SUBTRACTIVE METHOD

The problem is simple enough to be solved by asking, "What number added to 9 will equal 35?" Of course, the answer is 26.

CHECKING PROCESS

35 – 9 = 26, because 26 + 9 = 35

THE SHORT FORM

$$\begin{array}{r} 35 \\ \underline{-9} \end{array}$$

You cannot subtract 9 from 5.

To borrow 1 ten from 3 tens, draw a line through the 3 and write 2 above it, which represents the new minuend in that column.

$$\begin{array}{r} {}^{2}\not{3}5 \\ \underline{-9} \end{array}$$

The borrowed ten is exchanged for 10 ones and added to the 5 ones to get 15 ones.

To show this exchange, draw a line through the 5 and write 15 over it.

$$\begin{array}{r} {}^{2\ 15}\not{3}\not{5} \\ \underline{-9} \\ 26 \end{array}$$

9 from 15 = 6
0 from 2 = 2

Most students show the exchange by writing a 1 in front of the 5:

$$\begin{array}{r} \not{3}{}^{1}5 \\ \underline{-9} \\ 26 \end{array}$$

And there are many, many others who do the whole job mentally, like this:

$$\begin{array}{r} 35 \\ \underline{-9} \end{array}$$

9 from 15 = 6
0 from 2 = 2

PROOF: 26 + 9 = 35

TIME LIMIT–6 Minutes

SUBTRACTION

First time around, try using the SHORT-FORM method.
Note the combinations that cause you to hesitate.
The responses should not be thought over; they should be instinctive.
Second time around, try the ADDITIVE-SUBTRACTIVE method.

56	23	55	24	25	34	85	62	53	42
–9	–9	–6	–7	–8	–5	–8	–6	–6	–9

86	91	42	87	63	67	97	23	36	50
–7	–3	–4	–8	–4	–8	–9	–4	–7	–4

50	35	80	52	90	78	66	73	38	81
–3	–6	–6	–5	–2	–9	–9	–8	–9	–7

45	20	33	70	20	85	44	93	28	92
–8	–8	–7	–9	–1	–9	–6	–5	–9	–3

77	72	21	46	73	93	74	94	45	76
–8	–9	–3	–7	–5	–6	–8	–8	–6	–9

76	65	95	51	43	54	26	32	35	62
–8	–7	–8	–8	–8	–7	–7	–7	–8	–5

35	41	92	60	42	86	71	64	83	56
–9	–4	–8	–4	–6	–8	–7	–7	–7	–8

47	84	71	64	51	25	75	72	65	20
–9	–6	–5	–9	–7	–6	–7	–8	–9	–7

61	36	31	37	50	57	43	34	64	73
–2	–9	–7	–9	–7	–8	–7	–9	–6	–9

82	62	62	90	30	94	96	41	27	72
–8	–3	–7	–5	–8	–9	–8	–9	–9	–4

SCORING
1 to 4 errors = excellent
5 to 8 errors = good

A TWO-PLACE NUMBER FROM A TWO-PLACE NUMBER

Subtract

$$\begin{array}{r} 65 \\ -32 \\ \hline \end{array}$$

Mentally separate the problem into two columns.

$$\begin{array}{rcrr} 65 & = & 6 & 5 \\ -32 & = & -3 & -2 \\ \hline \end{array}$$

In this manner you reduce the problem to the solving of two basic subtraction facts.

2 from 5 = 3

3 from 6 = 3

PROOF:

65 – 32 = 33

33 + 32 = 65

Or you may reduce the problem to the solving of two basic addition facts.

"What number added to 2 equals 5?" = 3

"What number added to 3 equals 6?" = 6

NOTE: Problems like the above do not require BORROWING, therefore, they will never be any more difficult than finding the answer to questions that range from

"What number added to 1 equals 9?"

to

"What number added to 9 equals 9?"

Some students are daring and willing to practice methods that will produce answers in one big swoop, for example:

THE CRISS-CROSS METHOD

$$\begin{array}{r} 57 \\ -35 \\ \hline \end{array}$$

Procedure:

1) Take the full minuend (57).
2) Subtract the tens value of the subtrahend.

57 – 30 = 27

3) Subtract the ones value of the subtrahend.

27 – 5 = 22

Quick, wasn't it?–and all mental.

Let's try it again.

$$\begin{array}{r} 78 \\ -45 \\ \hline \end{array}$$

Think 78 – 40 = 38

38 – 5 = 33

$$\begin{array}{r} 89 \\ -57 \\ \hline \end{array}$$

Think 89 – 50 = 39

39 – 7 = 32

Don't forget the checking process.

89 – 57 = 32

because 32 + 57 = 89

TIME LIMIT—6 Minutes

SUBTRACTION

First time around, use the SHORT-FORM method.
Second time around, use the CRISS-CROSS method, for no reason other than as a mental exercise.
As you solve each problem, mentally check your accuracy.

59	78	37	46	97	28	69	85
−16	−24	−16	−11	−33	−17	−46	−62

55	79	38	47	96	27	68	89
−42	−18	−27	−22	−84	−11	−55	−71

97	77	39	45	57	76	39	46
−34	−42	−18	−23	−14	−63	−18	−22

78	95	66	58	49	88	76	29
−33	−54	−15	−42	−12	−56	−25	−17

54	76	37	43	97	88	66	77
−41	−32	−13	−20	−44	−31	−52	−13

79	38	59	65	47	88	99	28
−56	−10	−12	−43	−24	−55	−61	−12

34	57	89	68	39	45	56	88
−11	−25	−34	−12	−10	−21	−33	−62

SCORING

1 to 4 errors = excellent
5 to 8 errors = good

BORROWING

$$\begin{array}{r} 62 \\ -35 \\ \hline \end{array}$$

You cannot subtract 5 ones from 2 ones. You must borrow 1 ten from 6 tens, exchange it for 10 ones, and add it to 2 ones to get 12 ones.

The problem may be simplified by mentally separating it into two columns, like this:

$$\begin{array}{r} 62 \\ -35 \\ \hline \end{array} = \begin{array}{r} 5 \\ -3 \\ \hline \end{array} \quad \begin{array}{r} 12 \\ -5 \\ \hline \end{array}$$

In this manner you reduce the problem to the solving of two basic subtraction facts:

5 from 12 = 7

3 from 5 = 2

PROOF: 62 – 35 = 27

because 27 + 35 = 62

Or you may reduce the problem to the solving of two basic addition facts:

"What added to 5 will equal 12?" = 7

"What added to 3 will equal 5?" = 2

NOTE: Problems like the above do require BORROWING; however, they will never be any more difficult than finding the answer to questions that range from

"What added to 1 will equal 10?"

to

"What added to 9 will equal 18?"

Such problems are small enough to be solved by the mental exercise called

THE CRISS-CROSS METHOD

$$\begin{array}{r} 62 \\ -35 \\ \hline \end{array}$$

Think, 62 – 30 = 32

32 – 5 = 27

Problems with two-place numbers do not appear to give this method any trouble, whether they need borrowing or not.

A few more practice problems:

$$\begin{array}{r} 73 \\ -46 \\ \hline \end{array}$$

Think, 73 – 40 = 33

33 – 6 = 27

$$\begin{array}{r} 84 \\ -29 \\ \hline \end{array}$$

Think, 84 – 20 = 64

64 – 9 = 55

As always, don't forget the checking process.

84 – 29 = 55

because 55 + 29 = 84

TIME LIMIT—6 Minutes

SUBTRACTION

First time around, use the SHORT-FORM method.
Second time around, use the CRISS-CROSS method, for no reason other than as a mental exercise.
As you solve each problem, mentally check your accuracy.

56	74	36	41	93	27	66	82
−19	−28	−17	−16	−37	−18	−49	−65

52	78	37	42	94	21	65	81
−45	−19	−28	−27	−86	−17	−58	−79

94	72	38	43	54	73	38	42
−37	−47	−19	−25	−17	−66	−19	−26

73	94	65	52	42	86	75	27
−38	−55	−16	−48	−19	−58	−26	−19

51	72	33	40	94	81	62	73
−44	−36	−17	−23	−47	−38	−56	−17

76	30	52	63	44	85	91	22
−59	−18	−19	−45	−27	−58	−69	−18

31	55	84	62	30	41	53	82
−14	−27	−39	−18	−19	−25	−36	−68

SCORING

1 to 4 errors = excellent
5 to 8 errors = good

SUBTRACTING FROM A THREE-PLACE NUMBER

As the problems become progressively harder, all methods, except the Short-Form method, will be dropped, for obvious reasons.

EXAMPLE 1–No Borrowing

$$\begin{array}{r} 567 \\ -23 \\ \hline \end{array}$$

Mentally separate the problem into three columns.

Think,	3 from 7 =	4
	2 from 6 =	4
	0 from 5 =	5
	567 – 23 =	544

PROOF: 544 + 23 = 567

EXAMPLE 2–Borrowing From the Tens Column

$$\begin{array}{r} 355 \\ -29 \\ \hline \end{array}$$

You cannot subtract 9 ones from 5 ones. Borrow 1 ten from 5 tens, exchange it for 10 ones, and add to 5 ones to get 15 ones.

Problem regrouped:

$$\begin{array}{r} 3\overset{4}{\not{5}}\,\overset{1}{5} \\ -29 \\ \hline \end{array}$$

Think,	9 from 15 =	6
	2 from 4 =	2
	0 from 3 =	3
	355 – 29 =	326

PROOF: 326 + 29 = 355

EXAMPLE 3–Borrowing From the Hundreds Column

$$\begin{array}{r} 456 \\ -74 \\ \hline \end{array}$$

You cannot subtract 7 tens from 5 tens. Borrow 1 hundred from 4 hundreds, exchange it for 10 tens, and add to 5 tens to get 15 tens.

Problem regrouped:

$$\begin{array}{r} {}^{3}\not{4}\,\overset{1}{5}6 \\ -74 \\ \hline \end{array}$$

Think,	4 from 6 =	2
	7 from 15 =	8
	0 from 3 =	3
	456 – 74 =	382

PROOF: 382 + 74 = 456

EXAMPLE 4–Borrowing From Both the Tens and the Hundreds Columns

$$\begin{array}{r} 825 \\ -48 \\ \hline \end{array}$$

You cannot subtract 8 ones from 5 ones. Borrow 1 ten from 2 tens, exchange it for 10 ones, and add to 5 ones to get 15 ones.

Problem regrouped:

$$\begin{array}{r} 8\overset{1}{\not{2}}\,\overset{1}{5} \\ -48 \\ \hline \end{array}$$

You cannot subtract 4 tens from 1 ten. Borrow 1 hundred from 8 hundreds, exchange it for 10 tens, and add to 1 ten to get 11 tens.

Problem regrouped:

$$\begin{array}{r} {}^{7}\not{8}\,\overset{11}{\not{2}}\,\overset{1}{5} \\ -48 \\ \hline \end{array}$$

Think,	8 from 15 =	7
	4 from 11 =	7
	0 from 7 =	7
	825 – 48 =	777

PROOF: 777 + 48 = 825

TIME LIMIT—8 Minutes

SUBTRACTION

Use the SHORT-FORM method.
As you solve each problem, mentally check your accuracy.

Type 1								
	146	279	435	567	383	756	293	688
	−23	−34	−22	−45	−51	−24	−62	−56
	548	466	389	497	678	996	898	775
	−36	−42	−53	−62	−26	−73	−37	−14

Type 2								
	264	397	553	476	833	576	693	886
	−27	−49	−38	−57	−16	−48	−76	−69
	584	664	893	974	786	792	694	573
	−37	−28	−65	−49	−29	−56	−77	−66

Type 3								
	246	927	534	764	833	526	932	863
	−54	−43	−62	−81	−72	−93	−80	−70
	854	646	839	947	867	729	649	735
	−61	−75	−53	−82	−95	−47	−64	−74

Type 4								
	641	937	535	746	833	756	963	860
	−53	−59	−68	−77	−45	−89	−95	−72
	845	664	834	476	867	526	932	537
	−67	−95	−58	−88	−88	−79	−48	−69

SCORING

1 to 4 errors = excellent
5 to 8 errors = good

SUBTRACTING FROM THREE-PLACE NUMBERS (cont.))

EXAMPLE 5—No Borrowing

```
  476
 -235
```

Think, 5 from 6 = 1
3 from 7 = 4
2 from 4 = 2

476 − 235 = 241

PROOF: 241 + 235 = 476

EXAMPLE 6—Borrowing From the Tens Column

```
  476
 -237
```

Problem regrouped:

```
   6 1
  4 7 6
 -2 3 7
```

Think, 7 from 16 = 9
3 from 6 = 3
2 from 4 = 2

476 − 237 = 239

PROOF: 239 + 237 = 476

EXAMPLE 7—Borrowing From the Hundreds Column

```
  476
 -285
```

Problem regrouped:

```
  3 1
  4 7 6
 -2 8 5
```

Think, 5 from 6 = 1
8 from 17 = 9
2 from 3 = 1

476 − 285 = 191

PROOF: 191 + 285 = 476

EXAMPLE 8—Borrowing From Both the Tens and Hundreds Columns

```
  476
 -287
```

Problem regrouped:

```
   6 1
  4 7 6
 -2 8 7
```

Regrouped again:

```
  3 16 1
  4  7 6
 -2  8 7
```

Think, 7 from 16 = 9
8 from 16 = 8
2 from 3 = 1

476 − 287 = 189

PROOF: 189 + 287 = 476

TIME LIMIT–8 Minutes

SUBTRACTION

Use the SHORT-FORM method.
As you solve each problem, mentally check your accuracy.

Type 5	636	935	724	941	822	984	638	573
	–414	–622	–503	–120	–311	–361	–326	–252
	567	728	546	843	946	888	842	773
	–123	–614	–232	–623	–731	–617	–710	–602
Type 6	636	935	724	941	822	984	638	573
	–419	–628	–507	–126	–315	–366	–309	–255
	675	782	564	843	964	888	842	773
	–128	–614	–237	–625	–736	–619	–716	–608
Type 7	663	953	739	914	822	948	836	735
	–481	–672	–590	–150	–371	–361	–362	–252
	675	782	564	834	964	828	428	737
	–193	–490	–283	–651	–392	–666	–174	–373
Type 8	633	922	744	911	855	914	632	573
	–487	–637	–556	–228	–369	–737	–344	–296
	765	622	654	834	964	888	712	737
	–289	–356	–167	–248	–487	–299	–363	–459

SCORING

1 to 4 errors = excellent
5 to 8 errors = good

THE ZERO PROBLEM

A knowledge of the zero facts and the regrouping process is all that is necessary to solve any subtraction problem where zeros appear as digits.

ZERO FACTS

When a place in a numeral does not possess any value the digit 0 is used as a placeholder.

0 added to any number does not change the value of the number.

0 subtracted from any number does not change the value of the number.

The difference between two equal numbers is 0.

Any number subtracted from 0 becomes an unworkable problem.

EXAMPLE 1

$$\begin{array}{r} 320 \\ -6 \\ \hline \end{array}$$

You cannot subtract 6 ones from 0 ones. Borrow 1 ten from 2 tens, and exchange it for 10 ones.

Problem regrouped:

$$\begin{array}{r} 3\overset{1}{\not2}\overset{1}{0} \\ -6 \\ \hline \end{array}$$

6 from 10 = 4
0 from 1 = 1
0 from 3 = 3

320 – 6 = 314

PROOF: 314 + 6 = 320

EXAMPLE 2

$$\begin{array}{r} 306 \\ -8 \\ \hline \end{array}$$

You cannot subtract 8 ones from 6 ones. You cannot borrow any tens from 0 tens. Borrow 1 hundred from 3 hundreds, and exchange it for 10 tens.

Problem regrouped:

$$\begin{array}{r} \overset{2}{\not3}\overset{1}{0}6 \\ -8 \\ \hline \end{array}$$

Borrow 1 ten from 10 tens, exchange it for 10 ones, and add to 6 ones to get 16 ones.

Problem regrouped:

$$\begin{array}{r} \overset{2}{\not3}\overset{9}{\not1}\overset{1}{0}6 \\ -8 \\ \hline \end{array}$$

8 from 16 = 8
0 from 9 = 9
0 from 2 = 2

306 – 8 = 298

PROOF: 298 + 8 = 306

EXAMPLE 3

$$\begin{array}{r} 380 \\ -96 \\ \hline \end{array}$$

Borrow 1 ten from 8 tens, and exchange it for 10 ones.

Problem regrouped:

$$\begin{array}{r} 3\overset{7}{\not8}\overset{1}{0} \\ -96 \\ \hline \end{array}$$

Borrow 1 hundred from 3 hundreds, exchange it for 10 tens, and add to 7 tens to get 17 tens.

Problem regrouped:

$$\begin{array}{r} \overset{2}{\not3}\overset{17}{\not8}\overset{1}{0} \\ -96 \\ \hline \end{array}$$

6 from 10 = 4
9 from 17 = 8
0 from 2 = 2

380 – 96 = 284

PROOF: 284 + 96 = 380

TIME LIMIT–5 Minutes

SUBTRACTION

Use the SHORT-FORM method.
As you solve each problem, mentally check your accuracy.

40	50	60	30	80	90	70	20
–10	–30	–40	–20	–60	–30	–50	–20
43	56	67	33	88	92	75	28
–10	–30	–40	–20	–60	–30	–50	–20
408	506	607	309	805	906	707	208
–3	–5	–4	–6	–3	–3	–7	–6
403	505	604	306	803	903	707	206
–8	–6	–7	–9	–5	–6	–9	–8
480	560	670	390	850	960	770	280
–30	–50	–40	–60	–30	–30	–70	–60
430	550	640	360	830	930	720	260
–80	–60	–70	–90	–50	–60	–70	–80
480	560	670	390	850	960	770	280
–36	– 56	–44	–62	–37	–39	–78	–69
160	830	670	340	580	760	490	540
–37	–24	–39	–16	–69	–44	–78	–17

SCORING

1 to 4 errors = excellent
5 to 8 errors = good

EXAMPLE 4

```
  300
  -69
```

Borrow 1 hundred from 3 hundreds; exchange it for 10 tens.
Borrow 1 ten from 10 tens; exchange it for 10 ones.

Problem regrouped:

```
     9
   2 1̸ 1
   3̸0̸0
   -69
9 from 10 =   1
6 from  9 =   3
0 from  2 =   2
300 - 69 =  231
```

PROOF: 231 + 69 = 300

EXAMPLE 5

```
  306
 -108
```

Borrow 1 hundred from 3 hundreds; exchange it for 10 tens.
Borrow 1 ten from 10 tens, exchange it for 10 ones, and add to 6 ones to get 16 ones.

Problem regrouped:

```
     9
   2 1̸ 1
   3̸0̸6
  -108
8 from 16 =   8
0 from  9     9
1 from  2 =   1
306 - 108 = 198
```

PROOF: 198 + 108 = 306

EXAMPLE 6

```
  5,060
 -1,397
```

The regrouping process performed mentally:

```
      7 from 10 =      3
      9 from 15 =     6
      3 from  9 =    6
      1 from  4 =   3
5,060 - 1,397 = 3,663
```

PROOF: 3,663 + 1,397 = 5,060

EXAMPLE 7

```
  760,050
  -94,069
```

```
        9 from 10 =        1
        6 from 14 =       8
        0 from  9 =      9
        4 from  9 =     5
        9 from 15 =    6
        0 from  6 =   6
760,050 - 94,069 = 665,981
```

PROOF: 665,981 + 94,069 = 760,050

EXAMPLE 8

```
  3,630,009
 -1,708,500
```

```
            0 from  9 =          9
            0 from  0 =         0
            5 from 10 =        5
            8 from  9 =       1
            0 from  2 =      2
            7 from 16 =     9
            1 from  2 =   1
3,630,009 - 1,708,500 = 1,921,509
```

PROOF:
1,921,509 + 1,708,500 = 3,630,009

TIME LIMIT–6 Minutes

SUBTRACTION

Use the SHORT-FORM method.
As you solve each problem, mentally check your accuracy.

200	300	400	500	600	700	800	900
–30	–40	–20	–60	–80	–50	–70	–40
200	300	400	500	600	700	800	900
–31	–47	–26	–67	–85	–54	–73	–48
306	502	603	405	202	904	703	806
–30	–50	–60	–80	–40	–30	–40	–70
306	502	603	405	202	904	703	806
–38	–54	–67	–88	–49	–36	–47	–79
360	520	630	450	220	940	730	860
–80	–50	–60	–80	–40	–60	–40	–70
360	520	630	450	220	940	730	860
–83	–54	–67	–88	–49	–63	–47	–79
502	702	603	405	202	904	703	806
–205	–307	–409	–106	–108	–707	–508	–609
3,060	5,020	6,003	4,005	200,200	904,004	8,006,003	
–2,005	–3,070	–4,009	–1,060	–100,080	–700,707	–6,090,099	

SCORING

1 to 4 errors = excellent
5 to 8 errors = good

ESTIMATING DIFFERENCES

To estimate is to look at a problem and approximate the size of the answer.
To approximate means getting close to the correct answer.
Estimating should be done simply and quickly.
Complicated-looking numbers should be simplified.
Simplified numbers are numbers that have been rounded off to the nearest TEN, HUNDRED, and so forth.

EXAMPLE 1–Estimate the difference by rounding to the nearest ten.

	83	80
	−27	−30
Answer	56	
Estimate		50

EXAMPLE 2–Estimate the difference by rounding to the nearest hundred.

	625	600
	−273	−300
Answer	352	
Loose estimate		300

Estimate the difference by rounding to the nearest ten.

	625	630
	−273	−270
Answer	352	
Tight estimate		360

EXAMPLE 3–Estimate the difference by rounding to the nearest thousand.

	8,349	8,000
	−4,878	−5,000
Answer	3,671	
Loose estimate		3,000

Estimate the difference by rounding to the nearest hundred.

	8,349	8,300
	−4,878	−4,900
Answer	3,671	
Tight estimate		3,400

EXAMPLE 4–Estimate the difference by rounding to the nearest ten thousand.

	43,613	40,000
	−17,946	−20,000
Answer	25,667	
Loose estimate		20,000

Estimate the difference by rounding to the nearest thousand.

	43,613	44,000
	−17,946	−18,000
Answer	25,667	
Tight estimate		26,000

What kind of an estimate does one need?
It depends on the situation.
If you are estimating a money problem, a tight estimate is more useful.
If you are estimating a distance problem, a loose estimate may be all that is necessary.
For example: The distance to the sun is loosely 93,000,000 miles.

SUBTRACTION

573
–165

Estimated difference:
rounded to nearest hundred = ______
rounded to nearest ten = ______
Actual difference = ______

995
–622

Estimated difference:
rounded to nearest hundred = ______
rounded to nearest ten = ______
Actual difference = ______

6,723
–3,927

Estimated difference:
rounded to nearest thousand = ______
rounded to nearest hundred = ______
Actual difference = ______

9,263
–2,648

Estimated difference:
rounded to nearest thousand = ______
rounded to nearest hundred = ______
Actual difference = ______

89,452
–63,189

Estimated difference:
rounded to nearest ten thousand = ______
rounded to nearest thousand = ______
Actual difference = ______

341,960
–152,166

Estimated difference:
rounded to nearest hundred thousand = ______
rounded to nearest ten thousand = ______
Actual difference = ______

386
–124

Estimated difference:
rounded to nearest hundred = ______
rounded to nearest ten = ______
Actual difference = ______

854
–123

Estimated difference:
rounded to nearest hundred = ______
rounded to nearest ten = ______
Actual difference = ______

8,290
–5,634

Estimated difference:
rounded to nearest thousand = ______
rounded to nearest hundred = ______
Actual difference = ______

5,634
–1,592

Estimated difference:
rounded to nearest thousand = ______
rounded to nearest hundred = ______
Actual difference = ______

41,036
–17,877

Estimated difference:
rounded to nearest ten thousand = ______
rounded to nearest thousand = ______
Actual difference = ______

7,262,530
–3,807,642

Estimated difference:
rounded to nearest million = ______
rounded to nearest hundred thousand = ______
Actual difference = ______

TIME LIMIT—4 Minutes

ACHIEVEMENT TEST

37 −9	95 −19	56 −38	70 −22	60 −38
769 −527	473 −395	803 −389	592 −275	508 −345
2,375 −1,424	3,962 −3,026	8,702 −3,265	4,058 −2,497	5,859 −2,876
88,199 −61,726	86,028 −40,045	70,093 −29,728	43,228 −12,099	63,002 −47,983

72 − 39 =

417 − 75 =

3,921 − 77 =

385 − 128 =

4,227 − 1,886 =

TIME LIMIT–4 Minutes

REFRESHER TEST

73	52	45	37	95
68	73	77	95	43
62	62	86	72	68
43	37	93	63	78
97	78	67	23	73
84	76	87	32	69

6,657	3,652	2,399	6,623	9,454
8,344	5,236	6,372	9,463	4,587
5,076	3,218	4,589	3,345	1,551
9,687	7,144	5,374	3,053	9,186
4,684	5,839	5,468	5,654	6,964
8,419	3,794	5,947	7,575	3,987

RELATED PROBLEMS–SUBTRACTION

How does your checkbook balance out with the bank statement?

Balance brought forward		$750.00
Check #881		73.59
	Bank charge	.10
Balance		
Check #882		15.09
	Bank charge	.10
Balance		
Check #883		138.28
	Bank charge	.10
Balance		
Check #884		42.77
	Bank charge	.10
Balance		
Check #885		209.49
	Bank charge	.10
Balance		
AMOUNT DEPOSITED		427.59
Balance		
Check #886		88.26
	Bank charge	.10
Balance		

Balance brought forward		
Check #887		143.73
	Bank charge	.10
Balance		
Check #888		32.67
	Bank charge	.10
Balance		
Check #889		289.43
	Bank charge	.10
Balance		
Check #890		51.93
	Bank charge	.10
AMOUNT DEPOSITED		328.45
Balance		
Check #891		9.64
	Bank charge	.10
Balance		
Check #892		117.68
	Bank charge	.10
Balance		

The bank statement shows a balance of $292.28.

RELATED PROBLEMS—SUBTRACTION

1. Mr. and Mrs. Jones have decided to buy a home of their own.

The cost of the new house is	$29,750.00
As a down payment they will have to put up	4,462.50
What is the amount that will have to be mortgaged?	______

The bank notifies them that their monthly payments will amount to	$140.50
As of now, their rent amounts to	175.00
How much of a saving will they make?	______

Mr. Jones's take-home pay is	$687.75
How much will he have left after making the monthly mortgage payments?	______

2. The list price of a new car is $3,297.75
Mr. Brown bought it for 2,875.90
How much did he save? ______

3. Mrs. Stevens went shopping with $53.75
When she returned she had 18.69
How much did she spend? ______

4. The mileage on Mr. Brown's old car reads 43,284
Last year's reading was 35,729
How many miles did he travel last year? ______

5. Mr. Muller's weekly gross pay is $275.00

Mr. Muller's weekly gross pay is	$275.00
Federal income tax is	47.64
State income tax is	23.77
Social security tax is	16.89
City income tax is	1.77
Pension contribution is	13.09
NET PAY—Take-home pay is	______

Chapter 3

MULTIPLICATION

PRETEST–MULTIPLICATION

What is there about this operation that could cause someone like you to "climb a wall"?

How well do you know the basic facts?

Do you know that any number multiplied by 1 equals the number?

Do you know that any number multiplied by 0 equals 0?

Do you know that 4 x 3 and 3 x 4 provide the same answer?

When you carry a number, do you write it down–or do you leave it to memory and then forget it?

Do you estimate the answer as a check against the answer?

Do you have "zero" problems?

Do you set up the partial products properly?

Do you add the partial products properly?

Well, have we hit "home" yet?

If not, take the Pretest and let's find out if you're really honest with yourself–
we feel we've got a lot of interesting stuff
in this chapter–look it over.

PRETEST–MULTIPLICATION

218 5	952 7	858 8	905 5	850 9
5,080 7	6,009 8	7,600 6	58 60	70 49
86 35	90 80	860 47	907 39	805 40
608 504	800 938	560 476	895 70	370 560
5,080 63	6,900 78	4,795 49	4,090 708	8,400 590

MULTIPLICATION

If you buy 3 packs of gum and each pack contains 5 pieces, a total of how many pieces would you have in all?

You could arrive at the total in two ways:

By ADDITION $5 + 5 + 5 = 15$

By MULTIPLICATION $3 \times 5 = 15$

The total was arrived at by the repeated addition of the same addend.

Multiplication expresses the same problem as

3 groups of 5s = 15

The process of repeated addition is called MULTIPLICATION.

$3 \times 5 = 15$

The word "of" is replaced by the symbol "x".

The symbol "x" is used to indicate the operation of multiplication.

$3 \times 5 = 15$ may be read as:

3 groups of 5	equals 15
3 of 5	equals 15
3 5s	equals 15
5 multiplied by 3	equals 15
3 times 5	equals 15

The most popular way of expressing 3×5 is 3 times 5.

MULTIPLICATION PERFORMED ON THE NUMBER LINE

To multiply 5 by 3 we select a set of arrows each of equal length (in this case 5), and connect them in series like this;

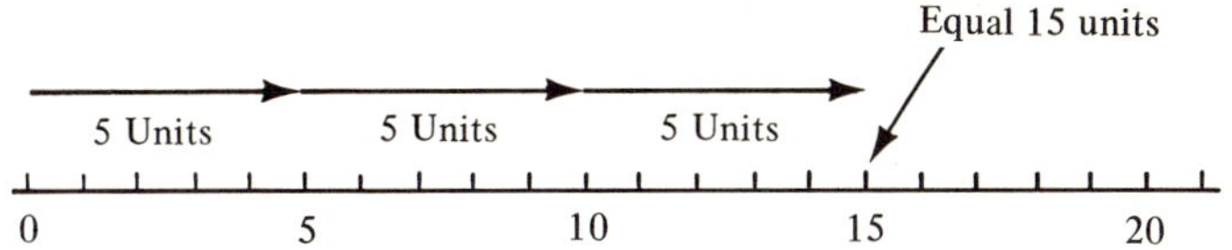

The number that represents the answer is located at the head of the last arrow.

Let's try it again.
This time instead of multiplying

3 times 5

let us multiply

5 times 3

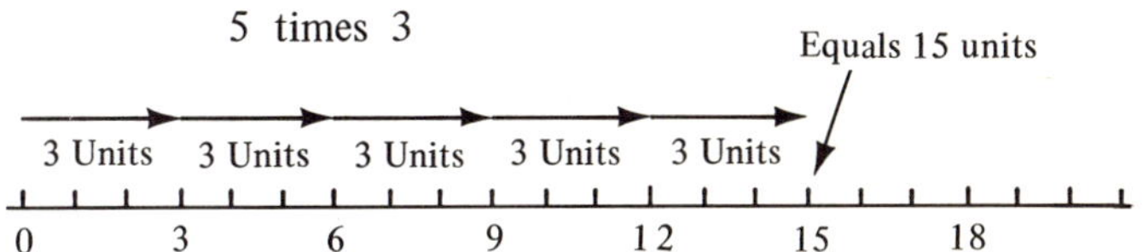

The head of the last arrow is resting over the number 15, which means:

5 times 3 equals 15

Strangely enough,

3 times 5 also equals 15

It becomes apparent that the COMMUTATIVE PROPERTY also holds for multiplication:

> You may multiply two numbers in any order without affecting the answer.

REMEMBER: This also holds true for ADDITION.

MULTIPLICATION TABLES

The repeated addition of the same addend introduces a new table where in one column an addend is chosen and in another column you are told the number of times the addend is used.

For example:

Number of times addend is used	Addend	Answer
1	2	2
2	2	4
3	2	6
4	2	8
5	2	10

The idea above is extended to include other addends. The following table includes all basic combinations.

Times Used	Addends 0	1	2	3	4	5	6	7	8	9
1	0	1	2	3	4	5	6	7	8	9
2	0	2	4	6	8	10	12	14	16	18
3	0	3	6	9	12	15	18	21	24	27
4	0	4	8	12	16	20	24	28	32	36
5	0	5	10	15	20	25	30	35	40	45
6	0	6	12	18	24	30	36	42	48	54
7	0	7	14	21	28	35	42	49	56	63
8	0	8	16	24	32	40	48	56	64	72
9	0	9	19	27	36	45	54	63	72	81

An example is presented to demonstrate how the table should be used to find the answer to:

5 times 4 = 20

OBSERVATION 1

The addend 0 repeated any number of times always produces the answer of 0.

PROOF: Why should 3 x 0 = 0?

3 x 0 means 3 groups of 0.

Rewritten in repeated addition form, we get

0 + 0 + 0 = 0

If we add nothing to nothing, we'll get nothing—true?

OBSERVATION 2

Any number multiplied by 1 always produces an answer that is the same as the number.

PROOF: Why should 1 x 3 = 3?

1 x 3 means 1 group of 3

which simply means 3 = 3

One 10-dollar bill is still 10 dollars—true?

OBSERVATION 3

2 x 3 = 6
3 x 2 = 6

1 x 3 = 3
3 x 1 = 3

0 x 5 = 0
5 x 0 = 0

Definite demonstration of the existence of the Commutative Property in Multiplication which states that:

You may multiply two numbers in any order without affecting the answer.

MULTIPLICATION—BASIC FACTS

TRAINING AID

Multiplication is a short way of adding equal numbers. Memorization of the basic multiplication facts is an absolute necessity.

REVIEW—Addition Table

2, 4, 6, 8, 10, 12, 14, 16, 18, 20

Solve mentally:

2 x 1 = 2	2 x 4 =	2 x 8 =
2 x 2 =	2 x 5 =	2 x 9 =
2 x 3 =	2 x 6 =	2 x 10 =
	2 x 7 =	

2 times 2 =	14 times 2 =	9 times 2 =
4 times 2 =	16 times 2 =	11 times 2 =
6 times 2 =	18 times 2 =	13 times 2 =
8 times 2 =	3 times 2 =	15 times 2 =
10 times 2 =	5 times 2 =	17 times 2 =
12 times 2 =	7 times 2 =	19 times 2 =

Multiply mentally:

8	18	5	15	3	13	6	16	7	17	9	19
x2	x2	x2	x2	x2	x2	x2	x2	x2	x2	x2	x2

2	2	2	2	2	2	2	2	2	2	2	2
x16	x6	x9	x19	x5	x15	x7	x17	x4	x14	x8	x18

Solve mentally:

2 multiplied by what number = 6? ________
2 multiplied by what number = 14? ________
2 multiplied by what number = 10? ________
2 multiplied by what number = 2? ________

What number multiplied by 2 = 22? ________
What number multiplied by 2 = 28? ________
What number multiplied by 2 = 32? ________
What number multiplied by 2 = 20? ________

2 multiplied by what number = 8? ________
2 multiplied by what number = 16? ________
2 multiplied by what number = 12? ________
2 multiplied by what number = 4? ________

What number multiplied by 2 = 30? ________
What number multiplied by 2 = 26? ________
What number multiplied by 2 = 34? ________
What number multiplied by 2 = 24? ________

MULTIPLICATION–BASIC FACTS

TRAINING AID

Multiplication is a short way of adding equal numbers. Memorization of the basic multiplication facts is an absolute necessity.

REVIEW–Addition Table

3, 6, 9, 12, 15, 18, 21, 24, 27, 30

Solve mentally:

3 x 1 =
3 x 2 =
3 x 3 =

3 x 4 =
3 x 5 =
3 x 6 =
3 x 7 =

3 x 8 =
3 x 9 =
3 x 10 =

2 times 3 =	14 times 3 =	9 times 3 =
4 times 3 =	16 times 3 =	11 times 3 =
6 times 3 =	18 times 3 =	13 times 3 =
8 times 3 =	3 times 3 =	15 times 3 =
10 times 3 =	5 times 3 =	17 times 3 =
12 times 3 =	7 times 3 =	19 times 3 =

Multiply mentally:

8	18	5	15	3	13	6	16	7	17	9	19
x3	x3	x3	x3	x3	x3	x3	x3	x3	x3	x3	x3

3	3	3	3	3	3	3	3	3	3	3	3
x16	x6	x9	x19	x5	x15	x7	x17	x4	x14	x8	x18

Solve mentally:

3 multiplied by what number = 6? ________
3 multiplied by what number = 15? ________
3 multiplied by what number = 12? ________
3 multiplied by what number = 3? ________

3 multiplied by what number = 9? ________
3 multiplied by what number = 18? ________
3 multiplied by what number = 21? ________
3 multiplied by what number = 27? ________

What number multiplied by 3 = 24? ________
What number multiplied by 3 = 33? ________
What number multiplied by 3 = 42? ________
What number multiplied by 3 = 30? ________

What number multiplied by 3 = 45? ________
What number multiplied by 3 = 36? ________
What number multiplied by 3 = 48? ________
What number multiplied by 3 = 39? ________

MULTIPLICATION–BASIC FACTS

TRAINING AID

Multiplication is a short way of adding equal numbers. Memorization of the basic multiplication facts is an absolute necessity.

REVIEW–Addition Table

4, 8, 12, 16, 20, 24, 28, 32, 40

Solve mentally:

4 x 1 =	4 x 4 =	4 x 8 =
4 x 2 =	4 x 5 =	4 x 9 =
4 x 3 =	4 x 6 =	4 x 10 =
	4 x 7 =	

2 times 4 =	14 times 4 =	9 times 4 =
4 times 4 =	16 times 4 =	11 times 4 =
6 times 4 =	18 times 4 =	13 times 4 =
8 times 4 =	3 times 4 =	15 times 4 =
10 times 4 =	5 times 4 =	17 times 4 =
12 times 4 =	7 times 4 =	19 times 4 =

Multiply mentally:

8	18	5	15	3	13	6	16	7	17	9	19
x4	x4	x4	x4	x4	x4	x4	x4	x4	x4	x4	x4

4	4	4	4	4	4	4	4	4	4	4	4
x16	x6	x9	x19	x5	x15	x7	x17	x4	x14	x8	x18

Solve mentally:

4 multiplied by what number = 8? ________
4 multiplied by what number = 16? ________
4 multiplied by what number = 12? ________
4 multiplied by what number = 4? ________

What number multiplied by 4 = 28? ________
What number multiplied by 4 = 44? ________
What number multiplied by 4 = 52? ________
What number multiplied by 4 = 60? ________

4 multiplied by what number = 20? ________
4 multiplied by what number = 32? ________
4 multiplied by what number = 24? ________
4 multiplied by what number = 36? ________

What number multiplied by 4 = 40? ________
What number multiplied by 4 = 48? ________
What number multiplied by 4 = 56? ________
What number multiplied by 4 = 64? ________

MULTIPLICATION—BASIC FACTS

TRAINING AID

Multiplication is a short way of adding equal numbers. Memorization of the basic multiplication facts is an absolute necessity.

REVIEW—Addition Table

5, 10, 15, 20, 25, 30, 35, 40, 45, 50

Solve mentally:

5 x 1 =	5 x 4 =	5 x 8 =
5 x 2 =	5 x 5 =	5 x 9 =
5 x 3 =	5 x 6 =	5 x 10 =
	5 x 7 =	

2 times 5 =	14 times 5 =	9 times 5 =
4 times 5 =	16 times 5 =	11 times 5 =
6 times 5 =	18 times 5 =	13 times 5 =
8 times 5 =	3 times 5 =	15 times 5 =
10 times 5 =	5 times 5 =	17 times 5 =
12 times 5 =	7 times 5 =	19 times 5 =

Multiply mentally:

8	18	5	15	3	13	6	16	7	17	9	19
x5	x5	x5	x5	x5	x5	x5	x5	x5	x5	x5	x5

5	5	5	5	5	5	5	5	5	5	5	5
x16	x6	x9	x19	x5	x15	x7	x17	x4	x14	x8	x18

Solve mentally:

5 multiplied by what number = 10? ________
5 multiplied by what number = 20? ________
5 multiplied by what number = 5? ________
5 multiplied by what number = 15? ________

What number multiplied by 5 = 60? ________
What number multiplied by 5 = 45? ________
What number multiplied by 5 = 50? ________
What number multiplied by 5 = 65? ________

5 multiplied by what number = 30? ________
5 multiplied by what number = 25? ________
5 multiplied by what number = 40? ________
5 multiplied by what number = 35? ________

What number multiplied by 5 = 55? ________
What number multiplied by 5 = 80? ________
What number multiplied by 5 = 70? ________
What number multiplied by 5 = 75? ________

MULTIPLICATION—BASIC FACTS

TRAINING AID

Multiplication is a short way of adding equal numbers. Memorization of the basic multiplication facts is an absolute necessity.

REVIEW—Addition Table

6, 12, 18, 24, 30, 36, 42, 48, 54, 60

Solve mentally:

6 x 1 =
6 x 2 =
6 x 3 =

6 x 4 =
6 x 5 =
6 x 6 =
6 x 7 =

6 x 8 =
6 x 9 =
6 x 10 =

2 times 6 =	14 times 6 =	9 times 6 =
4 times 6 =	16 times 6 =	11 times 6 =
6 times 6 =	18 times 6 =	13 times 6 =
8 times 6 =	3 times 6 =	15 times 6 =
10 times 6 =	5 times 6 =	17 times 6 =
12 times 6 =	7 times 6 =	19 times 6 =

Multiply mentally:

8	18	5	15	3	13	6	16	7	17	9	19
x6	x6	x6	x6	x6	x6	x6	x6	x6	x6	x6	x6

6	6	6	6	6	6	6	6	6	6	6	6
x16	x6	x9	x19	x5	x15	x7	x17	x4	x14	x8	x18

Solve mentally:

6 multiplied by what number = 36? ________
6 multiplied by what number = 12? ________
6 multiplied by what number = 54? ________
6 multiplied by what number = 6? ________

6 multiplied by what number = 30? ________
6 multiplied by what number = 42? ________
6 multiplied by what number = 18? ________
6 multiplied by what number = 60? ________

What number multiplied by 6 = 72? ________
What number multiplied by 6 = 24? ________
What number multiplied by 6 = 66? ________
What number multiplied by 6 = 96? ________

What number multiplied by 6 = 48? ________
What number multiplied by 6 = 78? ________
What number multiplied by 6 = 90? ________
What number multiplied by 6 = 84? ________

MULTIPLICATION—BASIC FACTS

TRAINING AID

Multiplication is a short way of adding equal numbers. Memorization of the basic multiplication facts is an absolute necessity.

REVIEW—Addition Table

7, 14, 21, 28, 35, 42, 49, 56, 63, 70

Solve mentally:

7 x 1 =	7 x 4 =	7 x 8 =
7 x 2 =	7 x 5 =	7 x 9 =
7 x 3 =	7 x 6 =	7 x 10 =
	7 x 7 =	

2 times 7 =	14 times 7 =	9 times 7 =
4 times 7 =	16 times 7 =	11 times 7 =
6 times 7 =	18 times 7 =	13 times 7 =
8 times 7 =	3 times 7 =	15 times 7 =
10 times 7 =	5 times 7 =	17 times 7 =
12 times 7 =	7 times 7 =	19 times 7 =

Multiply mentally:

8	18	5	15	3	13	6	16	7	17	9	19
x7	x7	x7	x7	x7	x7	x7	x7	x7	x7	x7	x7

7	7	7	7	7	7	7	7	7	7	7	7
x16	x6	x9	x19	x5	x15	x7	x17	x4	x14	x8	x18

Solve mentally:

7 multiplied by what number = 21? ________
7 multiplied by what number = 42? ________
7 multiplied by what number = 14? ________
7 multiplied by what number = 28? ________

7 multiplied by what number = 7? ________
7 multiplied by what number = 77? ________
7 multiplied by what number = 91? ________
7 multiplied by what number = 56? ________

What number multiplied by 7 = 49? ________
What number multiplied by 7 = 98? ________
What number multiplied by 7 = 56? ________
What number multiplied by 7 = 112? ________

What number multiplied by 7 = 70? ________
What number multiplied by 7 = 84? ________
What number multiplied by 7 = 35? ________
What number multiplied by 7 = 63? ________

MULTIPLICATION—BASIC FACTS

TRAINING AID

Multiplication is a short way of adding equal numbers. Memorization of the basic multiplication facts is an absolute necessity.

REVIEW—Addition Table

8, 16, 24, 32, 40, 48, 56, 64, 72, 80

Solve mentally:

8 x 1 =	8 x 4 =	8 x 8 =
8 x 2 =	8 x 5 =	8 x 9 =
8 x 3 =	8 x 6 =	8 x 10 =
	8 x 7 =	

2 times 8 =	14 times 8 =	9 times 8 =
4 times 8 =	16 times 8 =	11 times 8 =
6 times 8 =	18 times 8 =	13 times 8 =
8 times 8 =	3 times 8 =	15 times 8 =
10 times 8 =	5 times 8 =	17 times 8 =
12 times 8 =	7 times 8 =	19 times 8 =

Multiply mentally:

8	18	5	15	3	13	6	16	7	17	9	19
x8	x8	x8	x8	x8	x8	x8	x8	x8	x8	x8	x8

8	8	8	8	8	8	8	8	8	8	8	8
x16	x6	x9	x19	x5	x15	x7	x17	x4	x14	x8	x18

Solve mentally:

8 multiplied by what number = 24? ________
8 multiplied by what number = 48? ________
8 multiplied by what number = 16? ________
8 multiplied by what number = 32? ________

What number multiplied by 8 = 64? ________
What number multiplied by 8 = 128? ________
What number multiplied by 8 = 72? ________
What number multiplied by 8 = 104? ________

8 multiplied by what number = 40? ________
8 multiplied by what number = 80? ________
8 multiplied by what number = 8? ________
8 multiplied by what number = 56? ________

What number multiplied by 8 = 112? ________
What number multiplied by 8 = 88? ________
What number multiplied by 8 = 96? ________
What number multiplied by 8 = 120? ________

MULTIPLICATION–BASIC FACTS

TRAINING AID

Multiplication is a short way of adding equal numbers
Memorization of the basic multiplication facts is an
absolute necessity.

REVIEW–Addition Table

9, 18, 27, 36, 45, 54, 63, 72, 81, 90

Solve mentally:

9 x 1 =	9 x 4 =	9 x 8 =
9 x 2 =	9 x 5 =	9 x 9 =
9 x 3 =	9 x 6 =	9 x 10 =
	9 x 7 =	

2 times 9 =	14 times 9 =	9 times 9 =
4 times 9 =	16 times 9 =	11 times 9 =
6 times 9 =	18 times 9 =	13 times 9 =
8 times 9 =	3 times 9 =	15 times 9 =
10 times 9 =	5 times 9 =	17 times 9 =
12 times 9 =	7 times 9 =	19 times 9 =

Multiply mentally:

8	18	5	15	3	13	6	16	7	17	9	19
x9	x9	x9	x9	x9	x9	x9	x9	x9	x9	x9	x9

9	9	9	9	9	9	9	9	9	9	9	9
x16	x6	x9	x19	x5	x15	x7	x17	x4	x14	x8	x18

Solve mentally:

9 multiplied by what number = 27? ________
9 multiplied by what number = 54? ________
9 multiplied by what number = 18? ________
9 multiplied by what number = 36? ________

What number multiplied by 9 = 63? ________
What number multiplied by 9 = 126? ________
What number multiplied by 9 = 72? ________
What number multiplied by 9 = 144? ________

9 multiplied by what number = 45? ________
9 multiplied by what number = 90? ________
9 multiplied by what number = 9? ________
9 multiplied by what number = 99? ________

What number multiplied by 9 = 108? ________
What number multiplied by 9 = 135? ________
What number multiplied by 9 = 117? ________
What number multiplied by 9 = 81? ________

Multiplication may be presented horizontally like this:

4 x 6 = 24

or in a vertical form like this:

$$\begin{array}{r} 6 \\ \underline{x4} \\ 24 \end{array}$$

The 6 is called the MULTIPLICAND.
The 4 is called the MULTIPLIER.
The 24 is called the PRODUCT.

However, because

6 x 4

will produce the same product as

4 x 6

it has been decided that a common name should be used for both.

Therefore, all numbers involved in a multiplication operation are called <u>FACTORS</u>.

Combine together all the properties of multiplication and you will be able to reduce the memorization of the 90 Basic Multiplication facts down to 45.

1. You may multiply two factors in any order without affecting the product.
2. 0 multiplied by any factor equals 0.
3. Any factor multiplied by 1 equals the factor. Because multiplication by 1 does not change the identity of the factor being multiplied it is called the MULTIPLICATIVE IDENTITY.

NOTE: Examples of the Commutative Property in operation

0 1 0	1 1 1	2 1 2	3 1 3	4 1 4	5 1 5	6 1 6	7 1 7	8 1 8	9 1 9
0 2 0	1 2 2	2 2 4	3 2 6	4 2 8	5 2 10	6 2 12	7 2 14	8 2 16	9 2 18
0 3 0	1 3 3	2 3 6	3 3 9	4 3 12	5 3 15	6 3 18	7 3 21	8 3 24	9 3 27
0 4 0	1 4 4	2 4 8	3 4 12	4 4 16	5 4 20	6 4 24	7 4 28	8 4 32	9 4 36
0 5 0	1 5 5	2 5 10	3 5 15	4 5 20	5 5 25	6 5 30	7 5 35	8 5 40	9 5 45
0 6 0	1 6 6	2 6 12	3 6 18	4 6 24	5 6 30	6 6 36	7 6 42	8 6 48	9 6 54
0 7 0	1 7 7	2 7 14	3 7 21	4 7 28	5 7 35	6 7 42	7 7 49	8 7 56	9 7 63
0 8 0	1 8 8	2 8 16	3 8 24	4 8 32	5 8 40	6 8 48	7 8 56	8 8 64	9 8 72
0 9 0	1 9 9	2 9 18	3 9 27	4 9 36	5 9 45	6 9 54	7 9 63	8 9 72	9 9 81

TIME LIMIT—3 Minutes

MULTIPLICATION

All work should be mental, spontaneous and without hesitation.
Note the combinations that cause you to hesitate.

5 x3	7 x2	6 x1	4 x4	5 x4	3 x7	5 x3	7 x5	4 x3	5 x5
8 x6	4 x9	8 x3	2 x6	4 x4	2 x8	6 x5	3 x2	4 x7	1 x7
4 x6	9 x7	7 x6	7 x8	1 x5	8 x3	0 x5	4 x1	6 x3	5 x6
6 x2	1 x4	6 x7	3 x8	6 x9	0 x4	3 x3	5 x8	2 x6	0 x7
2 x5	4 x6	5 x3	5 x7	9 x5	7 x7	4 x8	4 x2	5 x4	6 x2
9 x6	7 x3	8 x1	9 x0	8 x4	8 x8	5 x8	3 x5	0 x6	9 x5
5 x9	8 x2	9 x3	1 x9	5 x4	7 x2	7 x4	7 x9	9 x3	1 x8
9 x8	4 x9	3 x8	2 x5	0 x4	6 x3	9 x1	9 x4	7 x8	5 x8

SCORING

1 to 4 errors = excellent
5 to 8 errors = good

A TWO-PLACE FACTOR BY A ONE-PLACE FACTOR

To solve such problems we shall use a process in which the memorization of the basic facts is all that is necessary.

For example:

$$\begin{array}{r} 23 \\ \underline{\times 3} \end{array}$$

EXPANDED NOTATION METHOD

This method is used only during the learning process—it permits us to look behind the operation.

$$\begin{array}{rcrcr} 2 \text{ tens} + 3 \text{ ones} & = & 20 + 3 & = & 23 \\ \underline{\times 3} & = & \underline{\times 3} & & \underline{\times 3} \\ 6 \text{ tens} + 9 \text{ ones} & = & 60 + 9 & & 69 \end{array}$$

The method above serves a very good purpose. It is used to cause you to recall that a number such as 23 means

2 tens + 3 ones

and if we have to multiply

2 tens + 3 ones by 3

it would be the same as adding

2 tens and 3 ones, 3 times

like this:

2 tens and 3 ones
2 tens and 3 ones
2 tens and 3 ones
6 tens and 9 ones = 69

NOTE: The method above distributed its operation to include addition—this property of multiplication is called the DISTRIBUTIVE PROPERTY.

THE SHORT-FORM METHOD

3 x 3 ones = 9 ones
Write 9 in the ones column.

$$\begin{array}{r} 23 \\ \underline{\times 3} \\ 9 \end{array}$$

3 x 2 tens = 6 tens
Write 6 in the tens column.

$$\begin{array}{r} 23 \\ \underline{\times 3} \\ 69 \end{array}$$

---

2 x 0 ones = 0 ones
Write 0 in the ones column.

$$\begin{array}{r} 30 \\ \underline{\times 2} \\ 0 \end{array}$$

2 x 3 tens = 6 tens
Write 6 in the tens column.

$$\begin{array}{r} 30 \\ \underline{\times 2} \\ 60 \end{array}$$

---

3 x 2 ones = 6 ones
Write 6 in the ones column.

$$\begin{array}{r} 52 \\ \underline{\times 3} \\ 6 \end{array}$$

3 x 5 tens = 15 tens
15 tens = 1 hundred and 5 tens

$$\begin{array}{r} 52 \\ \underline{\times 3} \\ 156 \end{array}$$

Write 5 in the tens column.
Write 1 in the hundreds column.

---

4 x 2 ones = 8 ones
Write 8 in the ones column.

$$\begin{array}{r} 52 \\ \underline{\times 4} \\ 8 \end{array}$$

4 x 5 tens = 20 tens
20 tens = 2 hundreds and 0 tens

$$\begin{array}{r} 52 \\ \underline{\times 4} \\ 208 \end{array}$$

Write 0 in the tens column.
Write 2 in the hundreds column.

TIME LIMIT–4 Minutes

MULTIPLICATION

Use the SHORT-FORM method

21 x4		82 x3		40 x4		43 x3		31 x4		93 x3		72 x4		13 x3	
	44 x2		31 x5		21 x6		73 x2		20 x5		31 x6		64 x2		21 x5
93 x3		52 x4		41 x7		30 x8		81 x6		40 x5		54 x2		12 x4	
	31 x6		82 x4		60 x5		51 x7		40 x8		93 x3		52 x4		21 x6
21 x7		52 x4		50 x8		71 x5		60 x9		52 x3		21 x6		10 x7	
	42 x4		83 x3		70 x5		61 x7		80 x8		74 x2		61 x9		30 x7
23 x3		71 x5		50 x9		82 x4		71 x7		30 x8		81 x7		70 x9	
	31 x5		50 x9		21 x7		63 x3		31 x8		64 x2		40 x7		90 x6

SCORING

1 to 4 errors = excellent
5 to 8 errors = good

CARRYING

What does one do when the multiplication of two one-place factors produces a two-place product? You cannot fit a two-place number into any one column.

For example:

```
 28
 x4
```

THE SHORT-FORM METHOD

All work must begin from the right.

```
 28
 x4
```

4 x 8 ones = 32 ones
32 ones = 3 tens + 2 ones
Write 2 in the ones column
and CARRY 3 to the
tens column, like this:

```
  3
 28
 x4
  2
```

4 x 2 tens = 8 tens
8 tens + "carried" 3 tens
= 11 tens
11 tens = 1 hundred + 1 ten
Write 1 in the tens column.
Write 1 in the hundreds column
like this:

```
  3
 28
 x4
112
```

REMEMBER: The carried number is always ADDED—never multiplied.

Now, if you could only carry in your head but let's not rush things—in due time you'll do it.

We prefer that you learn the Short-Form method but, if carrying is going to be a nuisance don't get upset—we have another method you can use. It is called the

PARTIAL-PRODUCT METHOD

```
 28
 x4
```

4 x 8 ones = 32 ones

32 ones = 3 tens + 2 ones
Write 2 in the ones column.
Write 3 in tens column
like this:

```
 28
 x4
 32
```

4 x 2 tens = 4 x 20 ones = 80 ones

80 ones = 8 tens + 0 ones
Write 8 in the tens column.
Write 0 in the ones column
like this:

```
 28
 x4
 32
 80
```

The sum of the partial products = 112

The Partial-Product method not only eliminates the need for carrying, but it also permits you to start the problem from the LEFT instead of from the RIGHT—another proof of the flexibility that exists in arithmetic

```
              28
              x4
4 x 20 =      80
4 x  8 =      32
             112
```

The Short-Form method is preferred—but if you have to use the Partial-Product method to get there—well, that's okay.

TIME LIMIT–6 Minutes

MULTIPLICATION

Solve the problems using the SHORT-FORM method
Carrying should be mental. If this is troublesome, then record the carried number or use the PARTIAL-PRODUCT method

24	85	47	44	38	95	74	16
x4	x3	x4	x3	x4	x3	x4	x3

48	33	24	78	25	34	67	26
x2	x5	x6	x2	x5	x6	x2	x5

96	58	47	33	85	52	59	17
x3	x4	x7	x8	x6	x5	x2	x4

36	85	64	53	42	96	57	23
x6	x4	x5	x7	x8	x3	x4	x6

27	56	55	74	64	59	28	47
x7	x4	x8	x5	x9	x3	x6	x7

48	87	76	65	84	79	62	33
x4	x3	x5	x7	x8	x2	x9	x7

28	77	53	86	75	34	84	74
x3	x5	x9	x4	x7	x8	x7	x9

38	54	25	67	33	66	45	93
x5	x9	x7	x3	x8	x2	x7	x6

SCORING

1 to 4 errors = excellent
5 to 8 errors = good

A THREE-PLACE FACTOR BY A ONE-PLACE FACTOR

Multiplication of larger factors is only an extended repetition of the procedures used on smaller problems.

SHORT-FORM METHOD–NO CARRYING

```
 123
  x2
   6
```

2 x 3 ones = 6 ones
Write 6 in the ones column.

```
 123
  x2
  46
```

2 x 2 tens = 4 tens
Write 4 in the tens column.

```
 123
  x2
 246
```

2 x 1 hundreds = 2 hundreds
Write 2 in the hundreds column.

Product = 246

SHORT-FORM METHOD–CARRYING

```
  2
 218
  x3
   4
```

3 x 8 ones = 24 ones
24 ones = 2 tens + 4 ones
Write 4 in the ones column.
Carry 2 to the tens column.

```
  2
 218
  x3
  54
```

3 x 1 tens = 3 tens
3 tens + carried 2 = 5 tens
Write 5 in the tens column.

```
  2
 218
  x3
 654
```

3 x 2 hundreds = 6 hundreds
Write 6 in the hundreds column.

REMEMBER–The carried number is always ADDED, never multiplied.

SHORT-FORM METHOD–CARRYING

```
  4
 329
  x5
   5
```

5 x 9 ones = 45 ones
45 ones = 4 tens + 5 ones
Write 5 in the ones column.
Carry 4 to the tens column.

```
 14
 329
  x5
  45
```

5 x 2 tens = 10 tens
10 tens + carried 4 = 14 tens
14 tens = 1 hundred + 4 tens
Write 4 in the tens column.
Carry 1 to the hundreds column.

```
   14
   329
    x5
 1,645
```

5 x 3 hundreds = 15 hundreds
15 hundreds + carried 1 = 16 hundreds
16 hundreds = 1 thousand + 6 hundreds
Write 6 in the hundreds column.
Write 1 in the thousands column.

Product = 1,645

PARTIAL-PRODUCT METHOD

To be used only if carrying is still a bit confusing.

```
 329
  x5
  45
```

5 x 9 ones = 45 ones
Write 5 in the ones column.
Write 4 in the tens column.

```
 329
  x5
  45
 100
```

5 x 20 = 100 ones
Write 0 in the ones column.
Write 0 in the tens column.
Write 1 in the hundreds column.

```
   329
    x5
    45
   100
 1,500
```

5 x 300 = 1,500 ones
Write 0 in the ones column.
Write 0 in the tens column.
Write 5 in the hundreds column.
Write 1 in the thousands column.

Sum of partial products = 1,645

TIME LIMIT—7 Minutes

MULTIPLICATION

Work all problems using the SHORT-FORM method. Carrying should be mental. If this is troublesome, then record the carried number or use the PARTIAL-PRODUCT method.

SCORING

1 to 4 errors = excellent
5 to 8 errors = good

361 x3	325 x5	193 x4	281 x6	436 x2	613 x7
234 x5	323 x6	426 x3	244 x8	157 x7	573 x4
386 x6	439 x3	918 x9	538 x7	384 x8	845 x8
457 x7	579 x4	798 x3	986 x2	867 x4	489 x5
672 x3	729 x4	293 x8	934 x5	342 x9	426 x6
268 x4	685 x6	852 x5	529 x8	295 x9	848 x3
657 x8	573 x4	739 x6	396 x5	964 x2	645 x4

THE ZERO PROBLEM

Some people think of 0 as being nothing; therefore they ignore it—how wrong they are! The digit 0, in a numeral, serves as a placeholder, filling in a place where no value exists. Without it our place-value system would collapse.

REMEMBER: The product of 0 and <u>any</u> factor equals 0.

EXAMPLE 1—Short-Form Method

8 x 0 ones = 0 ones
Write 0 in the ones column.

```
280
 x8
  0
```

8 x 8 tens = 64 tens
Write 4 in the tens column.
Carry 6 to the hundreds column.

```
6
280
 x8
 40
```

8 x 2 hundreds = 16 hundreds
16 hundreds + carried 6 = 22 hundreds
Write 2 in the hundreds column.
Write 2 in the thousands column.

```
  6
  280
   x8
2,240
```

Product = 2,240

EXAMPLE 2

7 x 9 ones = 63 ones
Write 3 in the ones column.
Carry 6 to the tens column.

```
 6
309
 x7
  3
```

7 x 0 tens = 0 tens
0 tens + carried 6 = 6 tens
Write 6 in the tens column.

```
 6
309
 x7
 63
```

7 x 3 hundreds = 21 hundreds
21 hundreds + no carry = 21 hundreds
Write 1 in the hundreds column.
Write 2 in the thousands column.

```
   6
  309
   x7
2,163
```

Product = 2,163

EXAMPLE 3

8 x 8 ones = 64 ones
Write 4 in the ones column.
Carry 6 to the tens column.

```
    6
2,008
   x8
    4
```

8 x 0 tens = 0 tens
0 tens + carried 6 = 6 tens
Write 6 in the tens column.

```
    6
2,008
   x8
   64
```

8 x 0 hundreds = 0 hundreds
0 hundreds + no carry = 0 hundreds
Write 0 in the hundreds column.

```
    6
2,008
   x8
  064
```

8 x 2 thousands = 16 thousands
16 thousands + no carry = 16 thousands
Write 6 in the thousands column.
Write 1 in the ten thousands column.

```
     6
 2,008
    x8
16,064
```

Product = 16,064

EXAMPLE 4

6 x 0 ones = 0 ones
Write 0 in the ones column.

```
4,090
   x6
    0
```

6 x 9 tens = 54 tens
Write 4 in the tens column.
Carry 5 to the hundreds column.

```
  5
4,090
   x6
   40
```

6 x 0 hundreds = 0 hundreds
0 hundreds + carried 5 = 5 hundreds
Write 5 in the hundreds column.

```
  5
4,090
   x6
  540
```

6 x 4 thousands = 24 thousands
24 thousands + no carry = 24 thousands
Write 4 in the thousands column.
Write 2 in the ten thousands column.

```
   5
 4,090
    x6
24,540
```

Product = 24,540

TIME LIMIT—7 Minutes

MULTIPLICATION

Use the SHORT-FORM method.
Because zeros can be troublesome, record the carrying.
Second time around, do the carrying mentally.
If the scoring is poor, use the PARTIAL-PRODUCT method.

SCORING

1 to 4 errors = excellent
5 to 8 errors = good

504	706	409	207	609	305
x8	x5	x2	x7	x6	x9
307	507	806	602	908	702
x4	x9	x3	x6	x5	x8
860	690	350	490	540	270
x3	x7	x5	x4	x6	x4
550	860	620	980	430	770
x7	x8	x7	x4	x5	x3
4,009	7,006	5,004	8,003	9,005	6,008
x6	x8	x4	x4	x7	x9
4,809	9,604	6,805	5,507	7,308	8,207
x8	x7	x8	x5	x6	x4
6,016	3,045	8,062	5,054	4,048	9,037
x5	x9	x7	x8	x4	x6
9,080	8,030	5,060	7,040	3,090	2,050
x3	x8	x6	x4	x5	x8

A TWO-PLACE FACTOR BY A TWO-PLACE FACTOR

You will recall that a number like 21 represents the sum of 20 and 1. Our problem is going to be solved by multiplying 43 first by 1 and then by 20, and adding the results.

```
 43
x21
```

SHORT-FORM METHOD

1 x 3 ones = 3 ones
Write 3 in the ones column.

```
 43
x21
  3
```

1 x 4 tens = 4 tens
Write 4 in the tens column.

```
 43
x21
 43
```

20 x 3 ones = 60 ones
60 ones = 6 tens + 0 ones

Write 0 in the ones column.
Write 6 in the tens column.

```
 43
x21
 43
 60
```

20 x 4 tens = 80 tens
80 tens = 8 hundreds
Write 8 in the hundreds column.

Sum of partial products =

```
 43
x21
 43
860
903
```

PARTIAL-PRODUCTS METHOD

1 x 3 =

1	x	3	=	3
1	x	40	=	40
20	x	3	=	60
20	x	40	=	800

```
 43
x21
  3
 40
 60
800
903
```

```
 43
x21
 43
860
903
```

SHORT-FORM METHOD

7 x 6 ones = 42 ones
Write 2 in the ones column.
Carry 4 tens.

```
  4
 36
x27
  2
```

7 x 3 tens = 21 tens
21 + carried 4 = 25 tens
25 tens = 2 hundreds + 5 tens
Write 5 in the tens column.
Write 2 in the hundreds column.

```
  4
 36
x27
252
```

20 x 6 ones = 120 ones
120 ones = 1 hundred + 2 tens
Write 0 in the ones column.
Write 2 in the tens column.
Carry 1 hundred.

```
  1
 36
x27
252
 20
```

20 x 3 tens = 60 tens
60 tens = 6 hundreds
6 + carried 1 = 7 hundreds
Write 7 in the hundreds column.
Product =

```
  1
 36
x27
252
720
972
```

BY THE WAY, did you know that the 0 in the second partial product may be left out? It is permitted only because "0 added to any number does not change the value of the number."

```
 36
x27
252
72
972
```

REMEMBER: If the end zero is to be left out, a space must be left for it.

PARTIAL-PRODUCTS METHOD

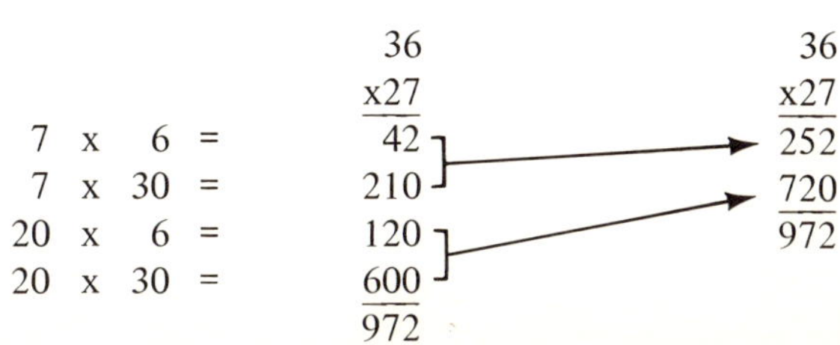

7	x	6	=	42
7	x	30	=	210
20	x	6	=	120
20	x	30	=	600

```
 36
x27
 42
210
120
600
972
```

```
 36
x27
252
720
972
```

TIME LIMIT—12 Minutes

MULTIPLICATION

SCORING

Use the SHORT-FORM method.
Take care placing the partial products.
Use end zeros to prevent errors.
Carrying should be performed mentally.

1 to 4 errors = excellent
5 to 8 errors = good

47	36	18	28	63	78	34	67
x31	x62	x44	x19	x37	x24	x46	x32

39	42	56	49	75	81	64	31
x17	x27	x41	x63	x36	x57	x49	x35

94	76	54	68	69	37	58	62
x24	x62	x18	x47	x29	x42	x22	x65

72	47	26	82	37	76	94	44
x67	x18	x56	x37	x21	x43	x72	x23

37	83	96	64	29	91	24	46
x64	x44	x23	x58	x63	x19	x38	x52

71	58	47	68	82	43	39	94
x24	x77	x36	x42	x25	x59	x64	x17

LARGE PROBLEMS

SHORT-FORM METHOD

First partial product

9 x 3 = 27
Write 7, carry 2.

```
   2
 273
x469
----
   7
```

NOTE: The first partial product begins in the same place-value position as the first multiplier digit 9.

9 x 7 = 63
63 + carried 2 = 65
Write 5, carry 6.

```
  62
 273
x469
----
  57
```

9 x 2 = 18
18 + carried 6 = 24
Write 24

```
  62
 273
x469
----
2457
```

Second partial product

6 x 3 = 18
Write 8, carry 1.

```
   1
 273
x469
----
2457
  8
```

NOTE: The second partial product begins in the same place-value position as the second multiplied digit 6.

6 x 7 = 42
42 + "1" = 43
Write 3, carry 4.

```
  41
 273
x469
----
2457
 38
```

6 x 2 = 12
12 + "4" = 16
Write 16.

```
  41
 273
x469
----
2457
1638
```

OBSERVATION

The end zero is absent from the second partial product.

Third partial product

4 x 3 = 12
Write 2, carry 1.

```
    1
  273
 x469
 ----
 2457
1638
  2
```

(NOTE: The third partial product begins in the same place-value position as the third multiplier digit 4.)

4 x 7 = 28
28 + "1" = 29
Write 9, carry 2.

```
   21
  273
 x469
 ----
 2457
1638
 92
```

4 x 2 = 8
8 + "2" = 10
Write 10.

```
    21
   273
  x469
  ----
  2457
 1638
1092
------
```

Sum of partial products = 128,037

The Partial-Product method is not recommended for large problems for obvious reasons—too many partial products.
Need convincing?

```
   273
  x469
  ----
    27
   630
  1800
   180
  4200
 12000
  1200
 28000
 80000
------
128,037
```

Convinced?

TIME LIMIT—20 Minutes | MULTIPLICATION | SCORING

Use the SHORT-FORM method. Take care placing the partial products.
If you have to use end zeros to prevent making errors, use them.
Carrying should be performed mentally.

1 to 2 errors = excellent
3 to 5 errors = good

863 x14	437 x25	762 x36	493 x43	743 x55
638 x19	347 x27	118 x28	326 x48	937 x56
817 x18	448 x28	927 x39	432 x45	183 x54
349 x168	815 x417	228 x356	473 x249	362 x454
595 x129	652 x737	734 x532	485 x125	564 x648

THE ZERO PROBLEM

EXAMPLE 1

First partial product

3 x 8 = 24
Write 4, carry 2.

3 x 0 = 0
0 + "2" = 2
Write 2.

3 x 4 = 12
Write 12.

```
   2
 408
x123
1224
```

Second partial product

2 x 8 = 16
Write 6, carry 1.

2 x 0 = 0
0 + "1" = 1
Write 1.

2 x 4 = 8
Write 8.

```
   1
 408
x123
1224
 816
```

Third partial product

1 x 8 = 8
Write 8.

1 x 0 = 0
Write 0.

1 x 4 = 4
Write 4.

```
   408
  x123
  1224
  816
 408
50,184
```

Sum partial products = 50,184

OBSERVATION: Each partial product begins in the same place-value position as its respective multiplier.

EXAMPLE 2

First partial product

3 x 0 = 0
Write 0.

3 x 8 = 24
Write 4, carry 2.

3 x 4 = 12
12 + "2" = 14
Write 14.

```
  2
 480
x103
1440
```

Second partial product

Any number multiplied by 0 = 0.

```
 480
x103
1440
 000
```

Third partial product

1 x 0 = 0
Write 0.

1 x 8 = 8
Write 8.

1 x 4 = 4
Write 4.

```
   480
  x103
  1440
  000
 480
49,440
```

Sum of partial products = 49,440

EXAMPLE 2– in abbreviated form

```
   480
  x103
  1440
 4800
49,440
```

It is not necessary to write out a string of zeros for the second partial product —one end zero is sufficient.

TIME LIMIT—15 Minutes

MULTIPLICATION

Use the SHORT-FORM method. Take care placing the partial products.
Try not using the end zeros.
Carrying should be performed mentally.

203	304	405	506	607	708	809	302	403
x234	x234	x234	x234	x234	x234	x234	x456	x456

504	605	706	807	908	230	340	450	560
x456	x456	x456	x456	x456	x204	x204	x204	x204

670	780	890	320	430	540	650	760	870	980
x204	x204	x204	x406	x406	x406	x406	x406	x406	x406

SCORING

1 to 2 errors = excellent
3 to 5 errors = good

THE ZERO PROBLEM (cont.)

EXAMPLE 3

First partial product	Any number multiplied by 0 = 0.	480 x130 000
Second partial product	3 x 0 = 0 Write 0. 3 x 8 = 24 Write 4, carry 2. 3 x 4 = 12 12 + "2" = 14 Write 14.	2 480 x130 000 1440
Third partial product	1 x 0 = 0 Write 0. 1 x 8 = 8 Write 8. 1 x 4 = 4 Write 4.	480 x130 000 1440 480 62,400

Sum of partial products = 62,400

EXAMPLE 3—in abbreviated form

480
x130
14400
480
62,400

One end zero is retained for the first partial product.

REMEMBER: All other partial products must begin in the same place–value position as the multiplier.

EXAMPLE 4

First partial product	Any number multiplied by 0 = 0.	408 x130 000
Second partial product	3 x 8 = 24 Write 4, carry 2. 3 x 0 = 0 0 + "2" = 2 Write 2. 3 x 4 = 12 Write 12	2 408 x130 000 1224
Third partial product	1 x 8 = 8 Write 8. 1 x 0 = 0 Write 0. 1 x 4 = 4 Write 4.	408 x130 000 1224 408 53,040

Sum of partial products = 53,040

EXAMPLE 4—in abbreviated form

408
x130
12240
408
53,040

One end zero is retained for the first partial product.

REMEMBER: All other partial products must begin in the same place-value position as the multiplier.

TIME LIMIT–15 Minutes

MULTIPLICATION

Use the SHORT-FORM method. Take care placing the partial products.
Try not using the end zeros.
Carrying should be performed.

703 x128	490 x352	607 x265	730 x434	806 x561	570 x610	905 x250	260 x560	
609 x340	390 x810	349 x307	493 x409	934 x607	699 x703	969 x806	817 x120	848 x350
927 x260	476 x430	783 x560	304 x309	590 x402	605 x706	730 x508	904 x605	

SCORING

1 to 2 errors = excellent
3 to 4 errors = good

ROUNDING OFF FACTORS AND ESTIMATING PRODUCTS

To estimate is to look at a problem and approximate the size of the answer. To approximate means getting close to the correct answer.

Estimating should be done quickly and simply.

Complicated-looking factors should be simplified. Simplified factors are factors that have been rounded off to the nearest TEN, HUNDRED, and so on.

ROUNDING OFF TO THE NEAREST TEN

47 should be rounded off to 50
because it is closer to
50 than it is to 40.

44 should be rounded off to 40
because it is closer to
40 than it is to 50.

247 should be rounded off to 250
because it is closer to
250 than it is to 240.

274 should be rounded off to 270
because it is closer to
270 than it is to 280.

ROUNDING OFF TO THE NEAREST HUNDRED

628 should be rounded off to 600
651 should be rounded off to 700

2,392 should be rounded off to 2,400
2,449 should be rounded off to 2,400

ROUNDING OFF TO THE NEAREST THOUSAND

3,456 should be rounded off to 3,000
9,975 should be rounded off to 10,000
12,455 should be rounded off to 12,000
249,544 should be rounded off to 250,000

SAMPLE EXERCISE

34,493 rounded to the nearest thousand 34,000
34,493 rounded to the nearest hundred 34,500
34,493 rounded to the nearest ten 34,490

IS 65 CLOSER TO 70 THAN IT IS TO 60?

RULE: If the second-place digit is 5 or greater than 5, INCREASE the first-place digit by 1.

65 should be rounded off to 70

RULE: If the second-place digit is 4 or less DO NOT INCREASE the first-place digit.

64 should be rounded off to 60

MULTIPLYING FACTORS WITH END ZEROS

10 x 5 = 50
Attach one zero to the product of 1 x 5.

20 x 40 = 800
Attach two zeros to the product of 2 x 4.

300 x 30 = 9,000
Attach three zeros to the product of 3 x 3.

400 x 400 = 160,000
Attach four zeros to the product of 4 x 4.

3,000 x 500 = 1,500,000
Attach five zeros to the product of 3 x 5.

6,000 x 4,000 = 24,000,000
Attach six zeros to the product of 6 x 4.

Keep your eye on this one:

50 x 200 = 10,000
Attach three zeros to the product of 5 x 2.

MULTIPLICATION

Round off to the nearest TEN:

134 ________

377 ________

728 ________

1,262 ________

Round off to the nearest HUNDRED:

134 ________

377 ________

728 ________

1,262 ________

Round off to the nearest THOUSAND:

412,345 ____________

847,633 ____________

3,516,717 ____________

7,632,500 ____________

3,340 has been rounded off to the nearest ____________________

3,400 has been rounded off to the nearest ____________________

3,000 has been rounded off to the nearest ____________________

73,000 has been rounded off to the nearest __________________

73,400 has been rounded off to the nearest __________________

73,410 has been rounded off to the nearest __________________

375,650 has been rounded off to the nearest _________________

375,600 has been rounded off to the nearest _________________

376,000 has been rounded off to the nearest _________________

20 x 3 = _______________

20 x 5 = _______________

30 x 10 = _______________

30 x 40 = _______________

400 x 3 = _______________

400 x 20 = _______________

400 x 30 = _______________

400 x 400 = _______________

5,000 x 5 = _______________

5,000 x 6 = _______________

5,000 x 30 = _______________

5,000 x 40 = _______________

5,000 x 200 = _______________

5,000 x 300 = _______________

5,000 x 5,000 = _______________

5,000 x 8,000 = _______________

ESTIMATING PRODUCTS

The work involves:

1. Rounding off factors
2. Product of basic combinations

EXAMPLE 1

Estimate the product of 46 x33

```
 46   rounded off to   50
x33   rounded off to  x30
138
138
1,518
```

Estimate: Attach two zeros to the product of
3 x 5 = 1,500

There's no question that the estimate of 1,500 is a very reasonable one.

OBSERVATION: Only one factor has a second-place number larger than 5.

EXAMPLE 2

Estimate the product of 575 x83

```
 575   rounded off to   600
 x83   rounded off to   x80
 1 725
46 00
47,725
```

Estimate: Attach three zeros to the product of
6 x 8 = 48,000

Once again we find the estimate to be a very reasonable one.

OBSERVATION: This problem too had only one factor with a second-place number larger than 5.

EXAMPLE 3

Estimate the product of 3,565 x1,613

OBSERVATION: Both factors have second-place numbers larger than 5.

```
If     3,565   is rounded off to    4,000
and   x1,613   is rounded off to   x2,000
      10 695
      35 65
    2 139 0
    3 565
   5,750,345
```

Estimate: Attach six zeros to the product of
4 x 2 = 8,000,000

We consider this estimate as being TOO HIGH.

```
If     3,565   is rounded off to    4,000
and   x1,613   is rounded off to   x1,000
   5,750,345
```

Estimate: Attach six zeros to the product of
4 x 1 = 4,000,000

We consider this estimate as being TOO LOW

```
If     3,565   is rounded off to    3,000
and   x1,613   is rounded off to   x2,000
   5,750,345
```

Estimate: Attach six zeros to the product of
3 x 2 = 6,000,000

We consider this estimate as being JUST RIGHT.

OBSERVATION: In the majority of cases when both factors have second place numbers larger than 5, you will be able to get a reasonable estimate by increasing the smaller factor.

TIME LIMIT– 25 Minutes

MULTIPLICATION

Round off the numbers.
Keep an eye on the digit with the second highest value.
Is the estimate reasonable?

	94 37	68 59	116 56	308 68	54,191 1,882
Actual	____	____	____	____	____
Estimate	____	____	____	____	____
	548 88	789 425	8,597 82	2,986 68	78,906 6,278
Actual	____	____	____	____	____
Estimate	____	____	____	____	____
	7,289 189	1,685 334	6,677 2,278	3,840 4,392	47,606 32,785
Actual	____	____	____	____	____
Estimate	____	____	____	____	____
	31,762 78	25,477 43	15,196 119	42,488 367	88,915 11,498
Actual	____	____	____	____	____
Estimate	____	____	____	____	____

SCORING

1 to 2 errors = excellent
3 to 4 errors = good

ACHIEVEMENT TEST 3

8 x6	7 x9	23 x7	37 x8	28 x7
64 x13	56 x36	94 x85	708 x6	209 x8
570 x5	405 x70	506 x40	690 x6	279 x83
794 x29	873 x195	347 x246	705 x457	409 x307
4,219 x6	5,785 x8	6,090 x9	5,008 x5	43,238 x7

Add:

505	597	6,147	26,157	188,599
607	286	7,402	93,318	313,607
809	318	3,086	64,007	269,396
		5,395	79,809	198,338

1,283	7,912	362	48
694	538	954	39
8,915	296	138	86
36	5,319	492	75
563	42	609	24
4,079	697	733	17
562		297	96
		328	53
		886	49
		575	68

Subtract:

439	513	729	408	1,209
–82	–88	–336	–272	–382

8,000	6,080	43,009	700,000	5,609,304
–4,070	–3,800	–28,072	–364,176	–3,937,169

Light travels at the rate of 186,000 miles per second.

How far will it travel in 1 minute? ________

How far will it travel in 1 hour? ________

How far will it travel in 1 day? ________

How far will it travel in 1 year (1 lightyear)? ________

One of our space ships traveled 28,335 miles
in one orbit about the earth.
One orbit took 2 hours.
How many miles did it travel in 1 day? ________

1 mile is equal to 5,280 feet.

How many feet in 24 miles? ________

1 kilometer is equal to 3,280 feet.

How many feet in 24 kilometers? ________

1 acre is equal to 43,560 square feet.

How many square feet in 13 acres? ________

1 cubic foot of water weighs approximately 62 pounds.

How many pounds in 18 cubic feet of water? ________

1,728 cubic inches equal 1 cubic foot.

How many cubic inches in 12 cubic feet? ________

27 cubic feet equal 1 cubic yard.

How many cubic feet in 15 cubic yards of concrete? ________

To pay for his new house,
Mr. Jones will have to make 360 monthly mortgage payments
of $140.50.
At the end of 360 months he will have paid the grand sum of: ______________

Of the $140.50, $70.25 will go against the principal,
28.68 will go for taxes, and
41.57 will go for interest.
After 360 payments,
tax payments will have amounted to: ______________
interest payments will have amounted to: ______________
principal payments will have amounted to: ______________

To carpet his new house, Mr. Jones bought 115 square yards of carpet
at $8.65 per square yard, installed.
The cost of the carpet is: ______________

For the new car he bought, he got himself a 24-month
bank loan. Each monthly payment was to be $67.25.
What was the full amount of the loan? ______________

Mr. Jones's weekly base salary is $275.00.
What does he gross in one year? ______________

Each week the following deductions are made from his pay:

Federal income tax	$47.64	In one year it amounts to:	______________
State income tax	23.77	In one year it amounts to:	______________
Social security	16.89	In one year it amounts to:	______________
City income tax	1.77	In one year it amounts to:	______________
Pension contribution	13.00	In one year it amounts to:	______________

His yearly take-home pay amounts to: ______________

His weekly budget calls for:

Food	$55.00	x	52	=	______________
Gas and electricity	3.35	x	52	=	______________
Insurance	2.72	x	52	=	______________
Car expense	15.25	x	52	=	______________
Clothing	15.20	x	52	=	______________
Entertainment	12.00	x	52	=	______________
House payments	32.43	x	52	=	______________
Miscellaneous	15.87	x	52	=	______________
Savings	20.00	x	52	=	______________
TOTAL		x	52	=	______________

Subtracted from his take-home pay, a balance of ______________ to spend as he wishes.

Chapter 4

DIVISION

PRETEST–DIVISION

Subtraction is an operation that has been analyzed as
"addition in reverse."

Multiplication is an operation that has been analyzed as
"repeated addition."

And now, DIVISION will be presented as an operation that will be analyzed as
"repeated subtraction."

It strikes us that all arithmetic operations are mere extensions of the addition operation.

And, if you're not good at addition, well–

Weakness in division may have as its chief cause:

Lack of mastery of the basic facts

Errors in multiplication

Errors in subtraction

Difficulty handling the remainder

Difficulty handling the zero problem

Estimating incorrectly

The pretest should reveal your weakness.

If one incorrect answer shows up, in a given row of examples, we may regard it as accidental.

However, if more than one error shows up, then some persistent difficulty exists.

The step-by-step approach used throughout this book will take you through the simplest skill and proceed to more complex skills until you reach that high level necessary to accomplish what you want.

6) 186	3) 1,860	3) 9,009	8) 584
9) 703	7) 5,469	4) 350	9) 6,000
5) 350	4) 1,632	7) 7,028	6) 6,402
77) 5,544	34) 13,804	48) 37,939	37) 2,892
35) 1,500	24) 4,678	53) 9,384	66) 67,002
34) 36,040	48) 196,320	78) 390,702	77) 539,539
609) 38,697	580) 38,692	636) 528,637	

DIVISION—BASIC FACTS

TRAINING AID

Division is defined as the inverse of multiplication.
You are given the product and one factor and are asked to find the other "factor."
Memorization of the basic multiplication facts is an absolute necessity.

REVIEW—Addition Table

2, 4, 6, 8, 10, 12, 14, 16, 18, 20

Solve mentally:

2 divided by 2 =	8 divided by 2 =	14 divided by 2 =
4 divided by 2 =	10 divided by 2 =	16 divided by 2 =
6 divided by 2 =	12 divided by 2 =	18 divided by 2 =

How many 2s are there in 18? ______
How many 2s are there in 12? ______
How many 2s are there in 22? ______

How many 2s are there in 14? ______
How many 2s are there in 24? ______
How many 2s are there in 16? ______

How many 2s are there in 20? ______
How many 2s are there in 28? ______
How many 2s are there in 30? ______

$\frac{8}{2} =$ $\frac{12}{2} =$ $\frac{10}{2} =$ $\frac{6}{2} =$ $\frac{14}{2} =$ $\frac{18}{2} =$ $\frac{14}{2} =$ $\frac{20}{2} =$ $\frac{16}{2} =$

$\frac{26}{2} =$ $\frac{30}{2} =$ $\frac{24}{2} =$ $\frac{36}{2} =$ $\frac{40}{2} =$ $\frac{28}{2} =$ $\frac{22}{2} =$ $\frac{32}{2} =$ $\frac{34}{2} =$

Solve mentally:

20 divided by 2 equals ______

14 divided by 2 equals ______

32 divided by 2 equals ______

44 divided by 2 equals ______

26 divided by 2 equals ______

18 divided by 2 equals ______

What number divided by 2 will equal 38? ______
What number divided by 2 will equal 24? ______
What number divided by 2 will equal 20? ______

What number divided by 2 will equal 30? ______
What number divided by 2 will equal 28? ______
What number divided by 2 will equal 14? ______

DIVISION–BASIC FACTS

TRAINING AID

Division is defined as the inverse of multiplication.
You are given the product and one factor and are asked to find the other "factor."
Memorization of the basic multiplication facts is an absolute necessity.

REVIEW–Addition Table

3, 6, 9, 12, 15, 18, 21, 24, 27, 30

Solve mentally:

6 divided by 3 = 15 divided by 3 = 24 divided by 3 =

9 divided by 3 = 18 divided by 3 = 27 divided by 3 =

12 divided by 3 = 24 divided by 3 = 30 divided by 3 =

How many 3s are there in 30? _____
How many 3s are there in 21? _____
How many 3s are there in 36? _____

How many 3s are there in 48? _____
How many 3s are there in 27? _____
How many 3s are there in 18? _____

How many 3s are there in 42? _____
How many 3s are there in 33? _____
How many 3s are there in 51? _____

$\frac{15}{3} =$ $\frac{12}{3} =$ $\frac{30}{3} =$ $\frac{27}{3} =$ $\frac{9}{3} =$ $\frac{18}{3} =$ $\frac{36}{3} =$ $\frac{24}{3} =$ $\frac{6}{3} =$

$\frac{21}{3} =$ $\frac{39}{3} =$ $\frac{48}{3} =$ $\frac{33}{3} =$ $\frac{51}{3} =$ $\frac{42}{3} =$ $\frac{3}{3} =$ $\frac{45}{3} =$ $\frac{54}{3} =$

Solve mentally:

18 divided by 3 equals _______
27 divided by 3 equals _______
33 divided by 3 equals _______

15 divided by 3 equals _______
24 divided by 3 equals _______
36 divided by 3 equals _______

What number divided by 3 will equal 42? _______
What number divided by 3 will equal 12? _______
What number divided by 3 will equal 9? _______

What number divided by 3 will equal 39? _______
What number divided by 3 will equal 48? _______
What number divided by 3 will equal 39? _______

DIVISION–BASIC FACTS

TRAINING AID

Division is defined as the inverse of multiplication.
You are given the product and one factor and are asked to find the other "factor."
Memorization of the basic multiplication facts is an absolute necessity.

REVIEW–Addition Table

4, 8, 12, 16, 20, 24, 28, 32, 36, 40

Solve mentally:

28 divided by 4 =	40 divided by 4 =	16 divided by 4 =
8 divided by 4 =	4 divided by 4 =	12 divided by 4 =
32 divided by 4 =	24 divided by 4 =	20 divided by 4 =

How many 4s are there in 48? ____
How many 4s are there in 24? ____
How many 4s are there in 16? ____

How many 4s are there in 32? ____
How many 4s are there in 28? ____
How many 4s are there in 20? ____

How many 4s are there in 12? ____
How many 4s are there in 52? ____
How many 4s are there in 60? ____

$\frac{20}{4} =$ $\frac{36}{4} =$ $\frac{12}{4} =$ $\frac{44}{4} =$ $\frac{60}{4} =$ $\frac{8}{4} =$ $\frac{68}{4} =$ $\frac{28}{4} =$ $\frac{52}{4} =$

$\frac{32}{4} =$ $\frac{4}{4} =$ $\frac{40}{4} =$ $\frac{64}{4} =$ $\frac{56}{4} =$ $\frac{72}{4} =$ $\frac{16}{4} =$ $\frac{48}{4} =$ $\frac{24}{4} =$

Solve mentally:

40 divided by 4 equals ____
12 divided by 4 equals ____
36 divided by 4 equals ____

28 divided by 4 equals ____
64 divided by 4 equals ____
32 divided by 4 equals ____

What number divided by 4 will equal 48? ____
What number divided by 4 will equal 40? ____
What number divided by 4 will equal 72? ____

What number divided by 4 will equal 16? ____
What number divided by 4 will equal 32? ____
What number divided by 4 will equal 60? ____

DIVISION—BASIC FACTS

TRAINING AID

Division is defined as the inverse of multiplication.
You are given the product and one factor and are asked to find the other "factor."
Memorization of the basic multiplication facts is an absolute necessity.

REVIEW—Addition Table

5, 10, 15, 20, 25, 30, 35, 40, 45, 50

Solve mentally:

55 divided by 5 =
25 divided by 5 =
30 divided by 5 =

35 divided by 5 =
20 divided by 5 =
60 divided by 5 =

15 divided by 5 =
45 divided by 5 =
65 divided by 5 =

How many 5s are there in 40? ______
How many 5s are there in 15? ______
How many 5s are there in 30? ______

How many 5s are there in 60? ______
How many 5s are there in 45? ______
How many 5s are there in 20? ______

How many 5s are there in 35? ______
How many 5s are there in 25? ______
How many 5s are there in 65? ______

$\frac{80}{5} =$ $\frac{40}{5} =$ $\frac{10}{5} =$ $\frac{70}{5} =$ $\frac{30}{5} =$ $\frac{60}{5} =$ $\frac{20}{5} =$ $\frac{50}{5} =$ $\frac{90}{5} =$

$\frac{15}{5} =$ $\frac{45}{5} =$ $\frac{65}{5} =$ $\frac{35}{5} =$ $\frac{5}{5} =$ $\frac{55}{5} =$ $\frac{75}{5} =$ $\frac{25}{5} =$ $\frac{85}{5} =$

Solve mentally:

60 divided by 5 equals ________
35 divided by 5 equals ________
20 divided by 5 equals ________

75 divided by 5 equals ________
45 divided by 5 equals ________
50 divided by 5 equals ________

What number divided by 5 will equal 30? ________
What number divided by 5 will equal 35? ________
What number divided by 5 will equal 90? ________

What number divided by 5 will equal 25? ________
What number divided by 5 will equal 60? ________
What number divided by 5 will equal 45? ________

DIVISION—BASIC FACTS

TRAINING AID

Division is defined as the inverse of multiplication.
You are given the product and one factor and are asked to find the other "factor."
Memorization of the basic multiplication facts is an absolute necessity.

REVIEW—Addition Table

6, 12, 18, 24, 30, 36, 42 48, 54 60

Solve mentally;

30 divided by 6 =	6 divided by 6 =	60 divided by 6 =
12 divided by 6 =	24 divided by 6 =	18 divided by 6 =
48 divided by 6 =	54 divided by 6 =	36 divided by 6 =

How many 6s are there in 48? _____
How many 6s are there in 24? _____
How many 6s are there in 12? _____

How many 6s are there in 60? _____
How many 6s are there in 30? _____
How many 6s are there in 90? _____

How many 6s are there in 72? _____
How many 6s are there in 36? _____
How many 6s are there in 18? _____

$\frac{30}{6} =$ $\frac{90}{6} =$ $\frac{18}{6} =$ $\frac{54}{6} =$ $\frac{78}{6} =$ $\frac{42}{6} =$ $\frac{12}{6} =$ $\frac{66}{6} =$ $\frac{102}{6} =$

$\frac{60}{6} =$ $\frac{24}{6} =$ $\frac{84}{6} =$ $\frac{48}{6} =$ $\frac{6}{6} =$ $\frac{96}{6} =$ $\frac{36}{6} =$ $\frac{72}{6} =$ $\frac{108}{6} =$

Solve mentally:

54 divided by 6 equals _______
42 divided by 6 equals _______
18 divided by 6 equals _______

90 divided by 6 equals _______
36 divided by 6 equals _______
18 divided by 6 equals _______

What number divided by 6 will equal 12? _______
What number divided by 6 will equal 24? _______
What number divided by 6 will equal 48? _______

What number divided by 6 will equal 18? _______
What number divided by 6 will equal 36? _______
What number divided by 6 will equal 72? _______

DIVISION–BASIC FACTS

TRAINING AID

Division is defined as the inverse of multiplication.
You are given the product and one factor and are asked to find the other "factor."
Memorization of the basic multiplication facts is an absolute necessity.

REVIEW–Addition Table

7, 14, 21, 28, 35, 42, 49, 56, 63, 70

Solve mentally:

42 divided by 7 =	35 divided by 7 =	21 divided by 7 =
14 divided by 7 =	7 divided by 7 =	28 divided by 7 =
48 divided by 7 =	56 divided by 7 =	63 divided by 7 =

How many 7s are there in 70? ______
How many 7s are there in 14? ______
How many 7s are there in 49? ______

How many 7s are there in 35? ______
How many 7s are there in 56? ______
How many 7s are there in 28? ______

How many 7s are there in 77? ______
How many 7s are there in 42? ______
How many 7s are there in 63? ______

$\frac{63}{7} =$ $\frac{21}{7}$ $\frac{49}{7} =$ $\frac{35}{7} =$ $\frac{7}{7} =$ $\frac{84}{7} =$ $\frac{98}{7} =$ $\frac{70}{7} =$ $\frac{112}{7} =$

$\frac{28}{7} =$ $\frac{105}{7} =$ $\frac{42}{7} =$ $\frac{14}{7} =$ $\frac{56}{7} =$ $\frac{119}{7} =$ $\frac{91}{7} =$ $\frac{126}{7} =$ $\frac{77}{7} =$

Solve mentally:

35 divided by 7 equals ________
28 divided by 7 equals ________
21 divided by 7 equals ________

70 divided by 7 equals ________
56 divided by 7 equals ________
42 divided by 7 equals ________

What number divided by 7 will equal 14? ________
What number divided by 7 will equal 49? ________
What number divided by 7 will equal 63? ________

What number divided by 7 will equal 84? ________
What number divided by 7 will equal 42? ________
What number divided by 7 will equal 21? ________

DIVISION–BASIC FACTS

TRAINING AID

Division is defined as the inverse of multiplication.
You are given the product and one factor and are asked to find the other "factor."
Memorization of the basic multiplication facts is an absolute necessity.

REVIEW–Addition Table

8, 16, 24, 32, 40, 48, 56, 64, 72, 80

Solve mentally;

16 divided by 8 =	24 divided by 8 =	40 divided by 8 =
32 divided by 8 =	48 divided by 8 =	80 divided by 8 =
64 divided by 8 =	72 divided by 8 =	120 divided by 8 =

How many 8s are there in 56? _____
How many 8s are there in 88? _____
How many 8s are there in 24? _____

How many 8s are there in 64? _____
How many 8s are there in 92? _____
How many 8s are there in 120? _____

How many 8s are there in 96? _____
How many 8s are there in 64? _____
How many 8s are there in 74? _____

$\frac{72}{8} =$ $\frac{24}{8} =$ $\frac{136}{8} =$ $\frac{88}{8} =$ $\frac{8}{8} =$ $\frac{56}{8} =$ $\frac{104}{8} =$ $\frac{120}{8} =$ $\frac{40}{8} =$

$\frac{64}{8} =$ $\frac{128}{8} =$ $\frac{16}{8} =$ $\frac{96}{8} =$ $\frac{48}{8} =$ $\frac{80}{8} =$ $\frac{32}{8} =$ $\frac{112}{8} =$ $\frac{144}{8} =$

Solve mentally:

56 divided by 8 equals _____
24 divided by 8 equals _____
32 divided by 8 equals _____

80 divided by 8 equals _____
40 divided by 8 equals _____
120 divided by 8 equals _____

What number divided by 8 will equal 16? _____
What number divided by 8 will equal 48? _____
What number divided by 8 will equal 96? _____

What number divided by 8 will equal 32? _____
What number divided by 8 will equal 64? _____
What number divided by 8 will equal 24? _____

DIVISION—BASIC FACTS

TRAINING AID

Division is defined as the inverse of multiplication.
You are given the product and one factor and are asked to find the other "factor."
Memorization of the basic multiplication facts is an absolute necessity.

REVIEW—Addition Table

9, 18, 27, 36, 45, 54, 63, 72, 81, 90

Solve mentally:

18 divided by 9 =

27 divided by 9 =

36 divided by 9 =

45 divided by 9 =

54 divided by 9 =

81 divided by 9 =

63 divided by 9 =

72 divided by 9 =

90 divided by 9 =

How many 9s are there in 36? _____
How many 9s are there in 99? _____
How many 9s are there in 71? _____

How many 9s are there in 54? _____
How many 9s are there in 90? _____
How many 9s are there in 63? _____

How many 9s are there in 99? _____
How many 9s are there in 81? _____
How many 9s are there in 27? _____

$\frac{27}{9} =$ $\frac{18}{9} =$ $\frac{63}{9} =$ $\frac{99}{9} =$ $\frac{45}{9} =$ $\frac{135}{9} =$ $\frac{81}{9} =$ $\frac{117}{9} =$ $\frac{9}{9} =$

$\frac{126}{9} =$ $\frac{36}{9} =$ $\frac{108}{9} =$ $\frac{144}{9} =$ $\frac{90}{9} =$ $\frac{162}{9} =$ $\frac{153}{9} =$ $\frac{72}{9} =$ $\frac{54}{9} =$

Solve mentally:

18 divided by 9 equals _______
36 divided by 9 equals _______
72 divided by 9 equals _______

27 divided by 9 equals _______
54 divided by 9 equals _______
81 divided by 9 equals _______

What number divided by 9 will equal 36? _______
What number divided by 9 will equal 108? _______
What number divided by 9 will equal 90? _______

What number divided by 9 will equal 72? _______
What number divided by 9 will equal 45? _______
What number divided by 9 will equal 135? _______

THE MEANING OF DIVISION

You can divide 12 dots into how many equal groups of 2s?

00/00/00/00/00/00 = 6 groups

You can divide 12 dots into how many equal groups of 3s?

000/000/000/000 = 4 groups

You can divide 12 dots into how many equal groups of 4s?

0000/0000/0000 = 3 groups

You can divide 12 dots into how many equal groups of 6s?

000000/000000 = 2 groups

You could find the number of equal groups in a given group by another process called REPEATED SUBTRACTION.

12	Group
−2	1
10	
−2	2
8	
−2	3
6	
−2	4
4	
−2	5
2	
−2	6
0	

12	Group
−3	1
9	
−3	2
6	
−3	3
3	
−3	4
0	

12	Group
−4	1
8	
−4	2
4	
−4	3
0	

12	Group
−6	1
6	
−6	2
0	

DIVISION, however, is a faster way of separating a given group into equal but smaller groups.

How many equal groups of 2s can you get out of a given group of 12? can be presented very simply as:

twelve divided by two,

which may be written as:

$12 \div 2, \quad \frac{12}{2}, \quad$ or $\quad 2\overline{)12}$

RELATED MULTIPLICATION AND DIVISION FACTS

16 divided by 4 = $\begin{array}{r} 4 \\ 4\overline{)16} \end{array}$

The answer is 4 because 4 x 4 = 16

30 divided by 5 = $\begin{array}{r} 6 \\ 5\overline{)30} \end{array}$

The answer is 6 because 6 x 5 = 30

48 divided by 6 = $\begin{array}{r} 8 \\ 6\overline{)48} \end{array}$

The answer is 8 because 8 x 6 = 48

OBSERVATION: Division and multiplication are closely related. Remembering the MULTIPLICATION FACTS alone should be sufficient for doing division.
You only have to remember that 8 x __ = 48 in order to answer 48 ÷ 8 = __?

THE MEANING OF DIVISION (cont.)

0 0 0 0
0 0 0 0
0 0 0 0

How many groups of 3s can you make? ________

How many groups of 4s can you make? ________

How many groups of 6s can you make? ________

0 0 0 0 0 0
0 0 0 0 0 0

How many groups of 2s can you make? ________

How many groups of 3s can you make? ________

How many groups of 4s can you make? ________

How many groups of 6s can you make? ________

0 0 0 0 0 0 0 0
0 0 0 0 0 0 0 0
0 0 0 0 0 0 0 0

How many groups of 2s can you make? ________

How many groups of 3s can you make? ________

How many groups of 4s can you make? ________

How many groups of 6s can you make? ________

How many groups of 8s can you make? ________

How many groups of 12s can you make? ________

A box containing 35 cookies is to be distributed among 5 boys.
How many cookies will each boy get? ________

There are 50 cigars in a box.
Mr. Smith smokes 5 cigars a day.
How many days will the box last him? ________

It took Jack 5 hours to drive a distance of 200 miles.
How many miles did he average each hour? ________

If	$7 \times 6 = 42$	then	$7\overline{)42}$ = ____
			$6\overline{)42}$ = ____
If	$8 \times 5 = 40$	then	$5\overline{)40}$ = ____
			$8\overline{)40}$ = ____
If	$6 \times 9 = 54$	then	$9\overline{)54}$ = ____
			$6\overline{)54}$ = ____
If	$9 \times 5 = 45$	then	$9\overline{)45}$ = ____
			$5\overline{)45}$ = ____
If	$3 \times 12 = 36$	then	$3\overline{)36}$ = ____
			$12\overline{)36}$ = ____
If	$4 \times 15 = 60$	then	$4\overline{)60}$ = ____
			$15\overline{)60}$ = ____
If	$15 \times 12 = 180$	then	$12\overline{)180}$ = ____
			$15\overline{)180}$ = ____

DIVISION—AN INVERSE OPERATION

In a multiplication operation we multiply two factors to get a product.

Factor	x	factor	=	product
4	x	5	=	20

In a division operation we divide the product by one of the factors to get the other factor.

$$\text{factor} \quad 4\overline{)20} = 5 \quad (5 \text{ factor, } 20 \text{ product})$$

The knowledge that division is the INVERSE of multiplication permits us to solve simple division problems such as:

$$3\overline{)15}$$

in the following manner:

What number multiplied by 3 will equal 15?

The answer, of course, is 5.

NOMENCLATURE

In a division problem you are asked to find, how many times a "given unit"
called the DIVISOR
is contained in a larger unit
called the DIVIDEND
The result you get is
called the QUOTIENT.

Division may be indicated by the sign ÷.
The problem $35 \div 5 = 7$

is read as 35 divided by 5 equals 7

35 is the DIVIDEND.
5 is the DIVISOR.
7 is the QUOTIENT.

Division may be indicated using the fraction form.

The problem $\frac{24}{4} = 6$

is read as 24 divided by 4 equals 6.

24 is the DIVIDEND.
4 is the DIVISOR.
6 is the QUOTIENT.

The problem $6\overline{)42}$ with quotient 7

is called the operational form.
It is read as 42 divided by 6 equals 7.

42 is the DIVIDEND.
6 is the DIVISOR.
7 is the QUOTIENT.

DIVISION BASIC FACTS

1. Any number divided by itself = 1.

$$\frac{5}{5} = 1 \qquad \frac{12}{12} = 1 \qquad \frac{250}{250} = 1 \qquad \frac{2{,}725}{2{,}725} = 1$$

2. Any number divided by 1 = the number.

$$\frac{5}{1} = 5 \qquad \frac{12}{1} = 12 \qquad \frac{250}{1} = 250 \qquad \frac{2{,}725}{1} = 2{,}725$$

3. Zero (0) divided by any number = 0.

$$\frac{0}{5} = 0 \qquad \frac{0}{12} = 0 \qquad \frac{0}{250} = 0 \qquad \frac{0}{2{,}725} = 0$$

4. Any number divided by 0 = NO SOLUTION
This is an unworkable problem.

$$\frac{5}{0} = \qquad \frac{12}{0} = \qquad \frac{250}{0} = \qquad \frac{2{,}725}{0} =$$

All are considered unworkable problems.

TIME LIMIT—3 Minutes

DIVISION

SCORING

All work should be mental, spontaneous and without hesitation.
Note the combinations that cause you to hesitate.

1 to 4 errors = excellent
5 to 8 errors = good

2) 12	9) 54	9) 27	5) 15	9) 72	7) 49
7) 56	8) 72	4) 36	7) 42	1) 9	3) 18
9) 54	7) 42	7) 28	8) 56	5) 40	6) 54
7) 63	8) 0	3) 18	9) 54	6) 36	3) 27
8) 64	1) 0	5) 20	5) 0	3) 27	1) 7
7) 28	0) 1	4) 32	2) 10	2) 0	2) 18
4) 0	6) 48	7) 49	0) 6	7) 63	5) 30
6) 42	4) 24	9) 45	8) 56	4) 36	0) 8

UNEVEN DIVISION

How many 3s are there in 12?

The answer, of course, is 4.

PROOF: $3 \times 4 = 12$

This is what is called EVEN DIVISION.

How many 3s are there in 13?

The answer is 4 with a REMAINDER of 1.

PROOF: $3 \times 4 + 1 = 13$

This is what is called UNEVEN DIVISION.

REMAINDERS-THEIR MEANING

000/000/000/000/0

Before you there are 13 dots.
They have been divided into groups of 3s.

However, you will observe that one of the groups is not equal in size to the others.

The 13 dots have been divided into 4 equal groups, with 1 dot left over.

The dot left over is called the REMAINDER.

00000/00000/00000/00000/0000

The above group contains 24 dots.
They have been divided into groups of 5s.

The 24 dots have been divided into 4 equal groups, with 4 dots left over.

The dots left over are called the REMAINDER.

UNEVEN DIVISIONS always yield a quotient with a REMAINDER that is always LESS than the divisor.

THE DIVISION PROCESS

The problem consists of a DIVIDEND of 13 that must be divided into equal groups of 3s. $3\overline{)13}$

We know that

$3 \times 1 = 3$
$3 \times 2 = 6$
$3 \times 3 = 9$
$3 \times 4 = 12$
$3 \times 5 = 15$

13 is not a product of the factor 3. The closest product of the factor 3 is 12. $3 \times 4 = 12$

1. How many 3s can you get out of 13? Only 4. $\begin{array}{r} 4 \\ 3\overline{)13} \end{array}$

2. Write 4 in the quotient above the 3.

   Why above the 3? Because the 13 represents 13 ones, and our 4 represents 4 ones, therefore, the proper place to record the 4 is in the ones place.

3. $4 \times 3 = 12$
   Write 12 under the 13 and subtract to get a REMAINDER of 1. $\begin{array}{r} 4 \\ 3\overline{)13} \\ \underline{12} \\ 1 \end{array}$

4. The QUOTIENT is 4 with a REMAINDER OF 1.
   It is written as 4 R1

5. PROOF: $4 \times 3 + 1 = 13$

OBSERVATION: In the checking process the REMAINDER is always ADDED.

TIME LIMIT–5 Minutes

DIVISION

Most problems can be worked out mentally.
However, do practice the DIVISION PROCESS method.
Check your work by adding the REMAINDER to the PRODUCT of the QUOTIENT and the DIVISOR.

$3\overline{)11}$ $3\overline{)17}$ $3\overline{)22}$ $3\overline{)29}$ $3\overline{)32}$ $3\overline{)38}$ $3\overline{)41}$ $3\overline{)44}$

$4\overline{)13}$ $4\overline{)18}$ $4\overline{)21}$ $4\overline{)25}$ $4\overline{)31}$ $4\overline{)38}$ $4\overline{)42}$ $4\overline{)49}$

$5\overline{)16}$ $5\overline{)13}$ $5\overline{)23}$ $5\overline{)28}$ $5\overline{)33}$ $5\overline{)39}$ $5\overline{)57}$ $5\overline{)64}$

$6\overline{)19}$ $6\overline{)15}$ $6\overline{)22}$ $6\overline{)27}$ $6\overline{)32}$ $6\overline{)37}$ $6\overline{)52}$ $6\overline{)67}$

$7\overline{)17}$ $7\overline{)20}$ $7\overline{)24}$ $7\overline{)29}$ $7\overline{)34}$ $7\overline{)43}$ $7\overline{)57}$ $7\overline{)62}$

$8\overline{)18}$ $8\overline{)20}$ $8\overline{)23}$ $8\overline{)26}$ $8\overline{)36}$ $8\overline{)41}$ $8\overline{)53}$ $8\overline{)61}$

$9\overline{)19}$ $9\overline{)14}$ $9\overline{)21}$ $9\overline{)28}$ $9\overline{)38}$ $9\overline{)48}$ $9\overline{)55}$ $9\overline{)70}$

SCORING

1 to 4 errors = excellent
5 to 8 errors = good

LARGER DIVIDENDS AND A ONE–PLACE DIVISOR

How many 3s in 80?

20? NO, because 20 x 3 = 60
30? NO, because 30 x 3 = 90

By these rough estimates we can conclude that the correct answer must be between 20 and 30. You could find the answer by continually subtracting 3 from 80, but that would take too much time. The division process is, of course, the operation to use: it is short and therefore a quick method.

Below you will find, side by side, two methods closely related. The first method will keep the steps in the process down to a simple level. The second method will take the same simple steps and condense them into what we call THE STANDARD DIVISION PROCESS.

METHOD 1–THE EXPANDED DIVISION PROCESS

1. How many 3s can you get out of 80?
   Are there 10? 20? 30?
   A good guess would be 20.
   Write 20 in the quotient.

$$3\overline{)80}$$

2. 20 x 3 = 60
   Write 60 below the 80 and subtract to get a REMAINDER of 20.

$$\begin{array}{r} 20 \\ 3\overline{)80} \\ \underline{60} \\ 20 \end{array}$$

3. <u>20 is the new dividend.</u>
   How many 3s can you get out of 20? Are there 5? 6? 7?
   A good guess would be 6.
   Write 6 in the quotient.

$$\begin{array}{r} 6 \\ 20 \\ 3\overline{)80} \\ \underline{60} \\ 20 \end{array}$$

4. 6 x 3 = 18
   Write 18 below the 20 and subtract to get a REMAINDER of 2.

5. How many 3s can you get out of 2? NONE.

$$\begin{array}{r} 6 \\ 20 \\ 3\overline{)80} \\ \underline{60} \\ 20 \\ \underline{18} \\ 2 \end{array}$$

6. The QUOTIENT is the sum of the partial quotients 20 + 6 = 26 with a REMAINDER of 2.

7. PROOF: 26 x 3 = 78 + 2 = 80

METHOD 2–THE STANDARD DIVISION PROCESS

1. How many 3s can you get out of 8? Only 2.
   Write 2 in the quotient above the 8.

$$3\overline{)80}$$

2. 2 x 3 = 6
   Write 6 below the 8 and subtract to get a REMAINDER of 2.

$$\begin{array}{r} 2 \\ 3\overline{)80} \\ \underline{6} \\ 2 \end{array}$$

   Bring down the 0 from the dividend and place it next to the 2.

3. <u>20 is the new dividend.</u>
   How many 3s can you get out of 20? Only 6.
   Write 6 in the quotient above the 0.

$$\begin{array}{r} 2 \\ 3\overline{)80} \\ \underline{6} \\ 20 \end{array}$$

4. 6 x 3 = 18
   Write 18 below the 20 and subtract to get a REMAINDER of 2.

$$\begin{array}{r} 26 \\ 3\overline{)80} \\ \underline{6} \\ 20 \\ \underline{18} \\ 2 \end{array}$$

5. How many 3s can you get out of 2? NONE.

6. The QUOTIENT is <u>26 R2</u>
   It is read as "26 with a remainder of 2"

7. PROOF: 26 x 3 + 2 = 80

TIME LIMIT—8 Minutes

DIVISION

Most problems can be worked out mentally.
However, do practice the DIVISION PROCESS method.
Check your work by adding the REMAINDER to the PRODUCT of the QUOTIENT and the DIVISOR.

8) 95	6) 77	3) 49	7) 81	8) 97	7) 92	
2) 53	5) 88	8) 103	9) 132	6) 94	4) 70	
5) 62	6) 70	4) 85	7) 89	9) 167	3) 70	
3) 76	4) 90	6) 100	7) 90	8) 100	9) 100	
4) 71	6) 82	7) 99	8) 129	5) 86	9) 122	
9) 130	6) 140	8) 130	7) 150	4) 130	3) 140	
7) 159	5) 116	3) 113	6) 176	8) 129	9) 139	

SCORING

1 to 4 errors = excellent
5 to 8 errors = good

LARGER DIVIDENDS AND A ONE-PLACE DIVISOR

Once again we present the EXPANDED FORM of division only because it helps to make each step in the STANDARD DIVISION PROCESS somewhat clearer. It has the advantage of showing in more detail how we arrive at the numerals which indicate the quotient. It has still another advantage in that it causes the student to work with "full partial quotients" at all times rather than parts of a "full final quotient." Once again we will demonstrate both methods side by side with the hope that it will lead you to "discover" and use the compact STANDARD DIVISION PROCESS.

METHOD 1–EXPANDED DIVISION PROCESS

1. How many 4s can you get out of 932? Are there 100? 200? 300? A good guess would be 200. Write 200 in the quotient.

```
   200
4 )932
```

2. 200 x 4 = 800
   Write 800 below the 932 and subtract to get a REMAINDER of 132.

```
   200
4 )932
   800
   ---
   132
```

3. 132 is the new dividend.
   How many 4s in 132? Are there 30? 40? 50? A good guess would be 30. Write 30 in the quotient.

4. 30 x 4 = 120
   Write 120 below the 132 and subtract to get a REMAINDER of 12.

```
    30
   200
4 )932
   800
   ---
   132
   120
   ---
    12
```

5. 12 is the new dividend.
   How many 4s in 12? Only 3. Write 3 in the the quotient.

6. 3 x 4 = 12
   Write 12 below the 12 and subtract to get a REMAINDER of 0.

```
     3
    30
   200
4 )932
   800
   ---
   132
   120
   ---
    12
    12
    --
     0
```

7. The QUOTIENT is the sum of the "partial quotients."
   200 + 30 + 3 = 233 + 0 = 233

8. PROOF: 233 x 4 + 0 = 932

METHOD 2–STANDARD DIVISION PROCESS

1. How many 4s can you get out of 9? Only 2. Write 2 in the quotient above the 9.

```
   2
4 )932
```

2. 2 x 4 = 8
   Write 8 below the 9 and subtract to get a REMAINDER of 1. Bring down the next digit in the dividend and place it beside the 1.

```
   2
4 )932
   8
   -
   13
```

3. 13 is the new dividend.
   How many 4s is 13? Only 3. Write 3 in the quotient above the 3.

4. 3 x 4 = 12
   Write 12 below the 13 and subtract to get a REMAINDER of 1. Bring down the next digit in the dividend and place it beside the 1.

```
   23
4 )932
   8
   -
   13
   12
   --
    1
```

5. 12 is the new dividend.
   How many 4s in 12? Only 3. Write 3 in the quotient above the 2.

6. 3 x 4 = 12
   Write 12 below the 12 and subtract to get a REMAINDER of 0.

```
   233
4 )932
   8
   -
   13
   12
   --
    12
    12
    --
     0
```

7. The QUOTIENT is 233 R0
   It is read as "233 with a remainder of 0."

8. PROOF: 233 x 4 + 0 = 932

TIME LIMIT–5 Minutes

DIVISION

Most problems can be worked out mentally.
However, do practice the DIVISION PROCESS method.
Check your work by adding the REMAINDER to the PRODUCT of the QUOTIENT and the DIVISOR.

2) 224 2) 247 2) 268 2) 439 2) 663 2) 851

3) 336 3) 364 3) 397 3) 645 3) 674 3) 938

4) 448 4) 487 4) 491 4) 812 4) 857 4) 889

5) 511 5) 572 5) 555 5) 537 5) 565 5) 593

6) 612 6) 636 6) 648 6) 666 6) 672 6) 694

7) 715 7) 749 7) 763 7) 777 7) 781 7) 798

8) 817 8) 832 8) 851 8) 864 8) 888 8) 899

SCORING

1 to 4 errors = excellent
5 to 8 errors = good

LARGER DIVIDENDS AND A ONE-PLACE DIVISOR

The next problem provides us with a dividend with a first digit NOT divisible by the divisor.

METHOD 1–EXPANDED DIVISION PROCESS

1. How many 7s can you get out of 329? Are there 10? 20? 30? 40? 50? A good guess would be 40. Write 40 in the quotient.

```
    40
7 ) 329
```

2. 40 x 7 = 280
   Write 280 below the 329 and subtract to get a REMAINDER of 49.

```
    40
7 ) 329
    280
     49
```

3. 49 is the new dividend.
   How many 7s in 49? Only 7.
   Write 7 in the quotient.

4. 7 x 7 = 49
   Write 49 below the 49 and subtract to get a REMAINDER of 0.

```
     7
    40
7 ) 329
    280
     49
     49
      0
```

5. The QUOTIENT is the sum of the partial quotients.

   40 + 7 = 47

6. PROOF: 47 x 7 = 329 + 0 = 329

METHOD 2–STANDARD DIVISION PROCESS

1. How many 7s in 3? NONE.
   Write 0 above the 3.

```
    0
7 ) 329
```

   How many 7s in 32? Only 4.
   Write 4 above the 2.

```
    04
7 ) 329
    28
     4
```

2. 4 x 7 = 28
   Write 28 below the 32 and subtract to get a REMAINDER of 4. Bring down the next digit in the dividend and place it beside the 4.

```
    04
7 ) 329
    28
     49
```

3. 49 is the new dividend.
   How many 7s in 49? Only 7.
   Write 7 above the 9.

4. 7 x 7 = 49
   Write 49 below the 49 and subtract to get a REMAINDER of 0.

```
    047
7 ) 329
    28
     49
     49
      0
```

5. The QUOTIENT is 047 R0
   It is read as "47 with a remainder of 0"

6. PROOF: 47 x 7 + 0 = 329

Do the numerals 047 and 47 represent the same value? YES.

A zero preceding a whole number does not change the value of the number.

WARNING: A zero following a whole number does change the value of the digit.

5 and 50 are not the same value, but 5 and 05 are the same.

TIME LIMIT–6 Minutes

DIVISION

Take care where you place the digits in the quotient.
If the first digit of the dividend is not divisible, place a zero above that digit.
Always check your answer.

7) 112	6) 210	7) 301	3) 132	5) 160	4) 327	5) 269
8) 176	9) 162	8) 128	9) 396	6) 198	7) 566	4) 339
7) 462	2) 196	8) 361	5) 442	9) 108	3) 277	6) 482
3) 162	2) 152	4) 164	3) 126	7) 322	8) 248	5) 417
6) 156	8) 144	4) 112	7) 105	6) 150	9) 615	5) 188
9) 405	9) 495	7) 252	3) 198	8) 112	6) 328	5) 227

SCORING

1 to 3 errors = excellent
4 to 6 errors = good

ESTIMATING QUOTIENTS

A skill in estimating answers has tremendous value.

1. It helps to prevent errors.
2. It gives you an idea what the size of the answer will be.

To estimate, round off the divisor and the dividend.

For example: 18) 793

If we round off the 18 to 20
and the 793 to 800,
the problem will look like this:

20) 800

The estimate is 40.

The correct quotient is 44 with a remainder of 1.

REMEMBER: The first digit of the divisor or dividend will remain unchanged if the digit with the second highest value is 4 or less.

The first digit of the divisor or dividend will be increased by 1 if the digit with the second highest value is 5 or more.

ESTIMATE 1: Estimate the quotient of

23) 1,259

23 is rounded off to 20
1,259 is rounded off to 1,000

The estimate is 50.

The correct quotient is 54 with a remainder of 7.

ESTIMATE 2: Estimate the quotient of 28) 8,752

28 is rounded off to 30
8,752 is rounded off to 9,000

The estimate is 300.

The correct quotient is 312 with a remainder of 16.

ESTIMATE 3: Estimate the quotient of

47) 6,247

47 is rounded off to 50
6,247 is rounded off to 6,000

The estimate is 120.

The correct quotient is 132 with a remainder of 43.

ESTIMATE 4: Estimate the quotient of

382) 8,217

382 is rounded off to 400
8,217 is rounded off to 8,000

The estimate is 20.

The correct quotient is 21 with a remainder of 195.

ESTIMATE 5: Estimate the quotient of

9,872) 436,496

9,872 is rounded off to 10,000
436,496 is rounded off to 400,000

The estimate is 40.

The correct quotient is 44 with a remainder of 2,128.

TIME LIMIT–6 Minutes

DIVISION

Do not work out the problems; all we want is an ESTIMATED QUOTIENT.
Round off the divisor and the quotient to the best of your ability.

$17\overline{)8{,}619}$ $28\overline{)8{,}512}$ $47\overline{)19{,}193}$ $64\overline{)32{,}128}$

$54\overline{)10{,}208}$ $18\overline{)19{,}833}$ $41\overline{)81{,}991}$ $32\overline{)26{,}159}$

$39\overline{)7{,}720}$ $62\overline{)117{,}336}$ $48\overline{)24{,}038}$ $23\overline{)43{,}618}$

$48\overline{)145{,}920}$ $78\overline{)718{,}560}$ $57\overline{)356{,}560}$ $36\overline{)715{,}380}$

$504\overline{)96{,}390}$ $859\overline{)717{,}453}$ $396\overline{)156{,}719}$ $618\overline{)64{,}916}$

$591\overline{)48{,}917}$ $609\overline{)35{,}697}$ $887\overline{)628{,}227}$ $714\overline{)486{,}347}$

$247\overline{)52{,}492}$ $722\overline{)83{,}721}$ $636\overline{)538{,}637}$ $147\overline{)501{,}359}$

SCORING

0 errors = excellent
1 to 2 errors = good

TWO–PLACE DIVISORS

METHOD 1–EXPANDED DIVISION PROCESS

1. How many 48s in 3,679? 48) 3,679
   To estimate:
   Round off 48 to 50
   Round off 3,679 to 4,000
   TRIAL QUOTIENT 80

2. 80 x 48 = ?
   8 x 8 = 64; write 4, carry 6.
   8 x 4 = 32 + 6 = 38

```
     80
48 ) 3,679
     3,840
```

3. 80 x 48 = 3,840
   The TRIAL QUOTIENT of 80 proves to be too large–let's try 70.

4. 70 x 48 = ?
   7 x 8 = 56; write 6, carry 5.
   7 x 4 = 28 + 5 = 33

5. 70 x 48 = 3,360–that's okay.

6. Write 70 in the quotient and write 3,360 below the 3,679 and subtract for a remainder of 319.

```
     70
48 ) 3,679
     3,360
       319
```

7. The new dividend is 319.
   How many 48s in 319?
   To estimate:
   Round off 48 to 50
   Round off 319 to 300
   TRIAL QUOTIENT 6

```
      6
     70
48 ) 3,679
     3,360
       319
       288
        31
```

8. 6 x 48 = ?
   6 x 8 = 48; write 8, carry 4.
   6 x 4 = 24 + 4 = 28

9. 6 x 48 = 288
   Write 6 in the quotient, write 288 below the 319, and subtract for a remainder of 31.

10. The quotient is 70 + 6 = 76 with a remainder of 31.

11. PROOF: 76 x 48 = 3,648 + 31 = 3,679

METHOD 2–STANDARD DIVISION PROCESS

1. How many 48s in 3? None.
   How many 48s in 36? None.
   How many 48s in 367?
   Round off 48 to 50
   Round off 367 to 370
   TRIAL QUOTIENT 7

```
48 ) 3,679
```

2. Write 7 above the 7.

3. 7 x 48 = ?
   7 x 8 = 56; write 6 and carry 5.
   7 x 4 = 28 + 5 = 33
   7 x 48 = 336

4. Write 336 below the 367 and subtract for a remainder of 31.

```
        7
48 ) 3,679
     3 36
       31
```

5. Bring down the last digit in the dividend and place it beside the 31.

```
        7
48 ) 3,679
     3 36
       319
```

6. The new dividend is 319.
   How many 48s in 319?
   Round off 48 to 50
   Round off 319 to 300
   TRIAL QUOTIENT 6

7. 6 x 48 = ?
   6 x 8 = 48; write 8 and carry 4.
   6 x 4 = 24 + 4 = 28
   6 x 48 = 288

```
       76
48 ) 3,679
     3 36
       319
       288
        31
```

8. Write 288 below the 319 and subtract for a remainder of 31.

9. The quotient is 76 R31

10. PROOF: 76 x 48 + 31 = 3,679

TIME LIMIT—20 Minutes

DIVISION

First, estimate the quotient.
Use the STANDARD DIVISION method.
Check your quotient against the estimate—was the estimate close?

$17\overline{)8{,}789}$ $28\overline{)8{,}792}$ $54\overline{)11{,}448}$ $33\overline{)20{,}163}$

$47\overline{)20{,}133}$ $91\overline{)83{,}831}$ $37\overline{)26{,}932}$ $64\overline{)33{,}460}$

$59\overline{)4{,}897}$ $48\overline{)2{,}544}$ $87\overline{)8{,}091}$ $77\overline{)5{,}621}$

$63\overline{)511{,}965}$ $29\overline{)148{,}667}$ $54\overline{)168{,}948}$ $87\overline{)619{,}685}$

$48\overline{)150{,}816}$ $78\overline{)711{,}594}$ $57\overline{)352{,}374}$ $36\overline{)261{,}108}$

SCORING

1 to 2 errors = excellent
3 to 4 errors = good

THE ZERO PROBLEM IN THE QUOTIENT

The chief difficulties involved in division have been rated in this order:

1. Failure to estimate the quotient properly
2. Mistakes in multiplication
3. Mistakes in subtraction
4. The handling of zero in the quotient

SAMPLE PROBLEM 1

```
31 ) 12,493
```

ESTIMATED QUOTIENT ---------------------------------- 400

1. How many 31s in 1? in 12? NONE.
   How many in 124? There are 4.

2. Write 4 above the digit 4.

```
       4
31 ) 12,493
     12 4
        0
```

3. 4 x 31 = 124
   Write 124 below the 124 and subtract for a remainder of 0.

4. Bring down the next digit (9) and place it beside the 0.

```
       4
31 ) 12,493
     12 4
        09
```

5. <u>The new dividend is 9.</u>
   How many 31s in 9? None.

THIS NEXT STEP IS AN ERROR SPOT

```
       40
31 ) 12,493
     12 4
        09
```

6. <u>Write 0 above the digit 9 as a placeholder.</u>

7. Bring down the next digit (3) and place it beside the 9.

```
       40
31 ) 12,493
     12 4
        093
```

8. <u>The new dividend is 93.</u>
   How many 31s in 93? There are 3.

9. Write 3 above the digit 3.

```
       403
31 ) 12,493
     12 4
        093
         93
          0
```

10. 3 x 31 = 93
    Write 93 below the 93 and subtract for a remainder of 0.

11. The quotient is 403.

12. PROOF: 403 x 31 = 12,493

SAMPLE PROBLEM 2

```
27 ) 6,210
```

ESTIMATED QUOTIENT -- 230

1. How many 27s in 6? NONE.
   How many in 62? There are 2.

2. Write 2 above the digit 2.

```
      2
27 ) 6,210
     5 4
       8
```

3. 2 x 27 = 54
   Write 54 below the 62 and subtract for a remainder of 8.

4. Bring down the next digit (1) and place it beside the 8.

```
      2
27 ) 6,210
     5 4
       81
```

5. <u>The new dividend is 81.</u>
   How many 27s in 81? There are 3.

6. Write 3 above the digit 1.

```
      23
27 ) 6,210
     5 4
       81
       81
        0
```

7. 3 x 27 = 81
   Write 81 below the 81 and subtract for a remainder of 0.

8. Bring down the next digit (0) and place it beside the 0.

```
      23
27 ) 6,210
     5 4
       81
       81
        00
```

9. <u>The new dividend is 00.</u>
   How many 27s in 00? NONE.

THIS NEXT STEP IS AN ERROR SPOT

10. <u>Write 0 above the digit 0 as a placeholder.</u>

```
      230
27 ) 6,210
     5 4
       81
       81
        00
```

11. The quotient is 230.

12. PROOF: 230 x 31 = 6,210

TIME LIMIT–10 Minutes

DIVISION

47) 4,935   64) 13,248   37) 29,896   29) 14,616

56) 33,936   78) 70,356   84) 25,872   97) 68,773

63) 9,450   42) 11,340   37) 2,960   71) 38,340

55) 36,300   86) 62,780   29) 25,230   97) 95,060

SCORING

1 error = excellent
2 to 3 errors = good

SAMPLE PROBLEM 3 — 33) 247,500

ESTIMATED QUOTIENT 8,000

1. How many 33s in 2? 24? NONE.
   How many in 247? There are 7.

2. Write 7 above the digit 7.

```
        7
33 ) 247,500
     231
      16
```

3. 7 x 33 = 231
   Write 231 below the 247 and subtract for a remainder of 16.

4. Bring down the next digit (5) and place it beside the 16.

```
        7
33 ) 247,500
     231
      16 5
```

5. The new dividend is 165.
   How many 33s in 165? There are 5.

6. Write 5 above the digit 5.

```
        7 5
33 ) 247,500
     231
      16 5
      16 5
         0
```

7. 5 x 33 = 165
   Write 165 below the 165 and subtract for a remainder of 0.

8. Bring down the next digit (0) and place it beside the 0.

```
        7 5
33 ) 247,500
     231
      16 5
      16 5
         00
```

9. The new dividend is 00.
   How many 33s in 00? NONE.

THIS NEXT STEP IS AN ERROR SPOT

10. Write 0 above the digit 0 as a placeholder.

```
        7 50
33 ) 247,500
     231
      16 5
      16 5
         00
```

11. Bring down the next digit (0) and place it beside the 00.

12. The new dividend is 000.
    How many 33s in 000? NONE.

THIS NEXT STEP IS AN ERROR SPOT

13. Write 0 above the second digit 0 as a placeholder.

```
        7 500
33 ) 247,500
     231
      16 5
      16 5
         000
```

14. The quotient is 7,500.

SAMPLE PROBLEM 4 — 44) 88,264

ESTIMATED QUOTIENT 2,000

1. How many 44s in 8? NONE.
   How many in 88? There are 2.

2. Write 2 above the second digit 8.

```
        2
44 ) 88,264
     88
      0
```

3. 2 x 44 = 88
   Write 88 below the 88 and subtract for a remainder of 0.

4. Bring down the next digit (2) and place it beside the 0.

```
        2
44 ) 88,264
     88
      0 2
```

5. The new dividend is 2.
   How many 44s in 2? NONE.

6. Write 0 above the digit 2 as a placeholder.

```
        2 0
44 ) 88,264
     88
      0 2
```

7. Bring down the next digit (6) and place it beside the 2.

```
        2 00
44 ) 88,264
     88
      0 26
```

8. The new dividend is 26.
   How many 44s in 26? NONE.

9. Write 0 above the digit 6 as a placeholder.

10. Bring down the next digit (4) and place it beside the 26.

```
        2,006
44 ) 88,264
     88
      0 264
        264
          0
```

11. The new dividend is 264.
    How many 44s in 264? There are 6.

12. Write 6 above the digit 4.

13. 6 x 44 = 264
    Write 264 below the 264 and subtract for a remainder of 0.

14. The quotient is 2,006.

TIME LIMIT–15 Minutes

DIVISION

First, estimate the quotient.
Use the STANDARD DIVISION method.
The quotients are loaded with zeros–take care.
Check the quotients against the estimates.

$36\overline{)61{,}200}$	$68\overline{)217{,}600}$	$47\overline{)216{,}200}$	$53\overline{)148{,}400}$
$74\overline{)429{,}200}$	$19\overline{)117{,}800}$	$94\overline{)836{,}600}$	$85\overline{)782{,}000}$
$34\overline{)36{,}040}$	$57\overline{)174{,}990}$	$48\overline{)196{,}320}$	$76\overline{)533{,}520}$
$46\overline{)46{,}276}$	$62\overline{)186{,}434}$	$78\overline{)390{,}702}$	$77\overline{)539{,}539}$

SCORING

1 error = excellent
2 to 3 errors = good

THREE-PLACE DIVISORS

Three-digit items are met daily in the business world.

For example: How many years are there in 78,284 days?
Let 1 year = 365 days.

```
365 ) 78,284
```

Estimate the quotient:
Round off 365 to 400
Round off 78,284 to....... 80,000
ESTIMATED QUOTIENT.....200

```
365 ) 78,284
```

1. How many 365s in 7? None.
   How many 365s in 78? None.
   How many 365s in 782?
   Round off 365 to 400
   Round off 782 to 800
   TRIAL QUOTIENT 2

   (Will 2 x 365 be less than 782? It looks okay.)

2. Write 2 above the digit 2.

```
          2
365 ) 78,284
```

3. 2 x 365 =
   2 x 5 = 10
   Write 0, carry 1.
   2 x 6 = 12 + 1 = 13
   Write 3, carry 1.
   2 x 3 = 6 + 1 = 7
   2 x 365 = 730

4. Write 730 below 782 and subtract for a remainder of 52.

```
          2
365 ) 78,284
      73 0
       5 2
```

5. Bring down the next digit (8) and place it beside the 52.

```
          2
365 ) 78,284
      73 0
       5 28
```

6. The new dividend is 528.
   How many 365s in 528? Only 1.

7. Write 1 above the digit 8.

8. 1 x 365 = 365

9. Write 365 below the 528 and subtract for a remainder of 163.

```
         21
365 ) 78,284
      73 0
       5 28
       3 65
       1 63
```

10. Bring down the last digit (4) and place it beside the 163.

11. The new dividend is 1,634.
    How many 365s in 1,634?
    Round off 365 to 400
    Round off 1,634 to 1,600
    TRIAL QUOTIENT.....................4

```
         21
365 ) 78,284
      73 0
       5 28
       3 65
       1,634
```

12. Write 4 above the digit 4.

13. 4 x 365 =
    4 x 5 = 20
    Write 0, carry 2.
    4 x 6 = 24 + 2 = 26
    Write 6, carry 2.
    4 x 3 = 12 + 2 = 14
    4 x 365 = 1,460

```
        214
365 ) 78,284
      73 0
       5 28
       3 65
       1,634
       1,460
         174
```

14. Write 1,460 below 1,634 and subtract for a remainder of 174.

15. The quotient is 214 R174

16. PROOF: 214 x 365 + 174 = 78,284

TIME LIMIT–25 Minutes

DIVISION

First, estimate the quotient.
Check your answers against the estimates–do they look reasonable?

$504\overline{)96,390}$ $859\overline{)697,453}$ $396\overline{)156,719}$ $618\overline{)64,916}$

$591\overline{)48,917}$ $609\overline{)38,697}$ $887\overline{)680,227}$ $714\overline{)448,347}$

$965\overline{)55,753}$ $219\overline{)548,517}$ $908\overline{)470,743}$ $198\overline{)109,700}$

$615\overline{)320,975}$ $580\overline{)38,692}$ $922\overline{)293,529}$ $309\overline{)985,257}$

$247\overline{)62,492}$ $722\overline{)88,721}$ $636\overline{)528,637}$ $147\overline{)501,359}$

SCORING

1 error = excellent
2 errors = good

TIME LIMIT—20 Minutes

ACHIEVEMENT TEST 4

$7\overline{)1{,}407}$	$8\overline{)280}$	$6\overline{)1{,}188}$	$9\overline{)635}$
$8\overline{)7{,}000}$	$23\overline{)414}$	$36\overline{)838}$	$62\overline{)2{,}728}$
$77\overline{)4{,}235}$	$80\overline{)7{,}000}$	$47\overline{)573{,}619}$	$28\overline{)417{,}525}$
$39\overline{)313{,}632}$	$54\overline{)269{,}513}$	$68\overline{)497{,}737}$	$516\overline{)40{,}250}$
$368\overline{)23{,}555}$	$233\overline{)717{,}640}$	$339\overline{)1{,}425{,}495}$	$407\overline{)842{,}089}$

TIME LIMIT–8 Minutes

REFRESHER TEST 3

Add:

536	305	78	8,668	38,073
757	609	67	485	567
779	776	60	879	4,439
963	308	96		856,628
366	507	69		

Subtract:

128,106	236,464	16,046	62,020	153,103
–68,759	–47,865	–9,798	–61,096	–93,756

Multiply:

6,070	3,395	8,008	6,060	7,506
x83	x57	x267	x394	x408

If the distance to and from the moon, plus several orbits around the planet, equals a total distance of 624,960 miles, and a trip took 12 full days, what was the average distance traveled each day? ____________

What was the average hourly speed of the space ship? ____________

How much money will each of 33 stockholders receive if they are to divide $35,640 equally? ____________

There are 27 cubic feet in 1 cubic yard. To pave a terrace, it has been estimated that a total of 351 cubic feet of concrete will be needed.

How many cubic yards should be ordered? ____________

There are 9 square feet to a square yard.
It has been estimated that a total of 405 square feet of linoleum should be ordered.

Convert this order into square yards. ____________

There are 36 inches to a yard.

How many yards in 1,620 inches? ____________

1,760 yards equal 1 mile.

How many miles in 49,280 yards? ____________

3,280 feet equal 1 kilometer.

How many kilometers in 508,400 feet? ____________

5,280 feet equal 1 mile.

How many miles in 506,880 feet? ____________

1,728 cubic inches equal 1 cubic foot.

How many cubic feet in 181,440 cubic inches? ____________

16 ounces equal 1 pound.

How many pounds in 19,296 ounces? ____________

Mr. Crown borrowed $3,888 from the bank.
He must repay in 36 months.
What will be his monthly payment? __________

A man earned $213.85 for a 5-day week.
What is his average daily wage? __________
If 35 hours constitute a workweek, what is his hourly rate? __________

Mrs. Thayer bought 28 square yards of carpeting for $504. What was the cost per square yard? __________

There are 36 inches to a yard.
How many yards in 2,592 inches? __________

There are 5,280 feet to a mile.
How many miles in 554,400 feet? __________

A racing car traveled 1,416 miles in 12 hours.
What was the average speed? __________

Over a period of 2 years, Joan saved $1,872.
How much did she save each month? __________
How much did she save each week? __________

A trip touring the U.S. will cover 2,520 miles.

If the car can cover 14 miles per gallon, how many gallons will have to be bought? __________

If the trip is to be made in 8 days, how many miles should be traveled each day? __________

If we travel 360 miles each day, the trip will take how many days? __________

Chapter 5

FRACTIONS

PRETEST–FRACTIONS

No other number dishes out more trouble than the "little old fraction."

For some reason or other, people do not believe that fractions are subject to the very same basic laws and rules that apply to whole numbers.

Consider this example:

> One fine day Johnny did some work for his mother, who promised him 1/2 of a pie for his effort.
>
> On the next day he did a little more work and this time he was promised 1/3 of a pie.
>
> When it came time for him to receive his pay, his mother offered him: $\frac{1}{2} + \frac{1}{3} = \frac{2}{5}$ of a pie as full payment.
>
> Johnny accepted his payment but with a skeptical look on his face. He figured something was wrong.
>
> Now, what was wrong?

Visually, this is the transaction his mother worked out;

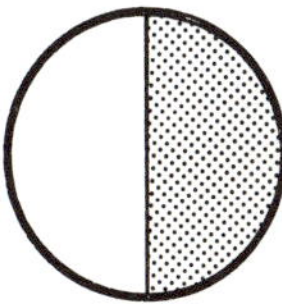

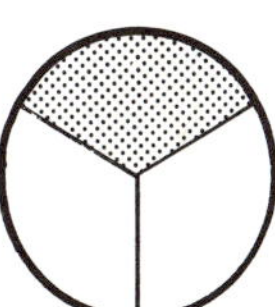

=

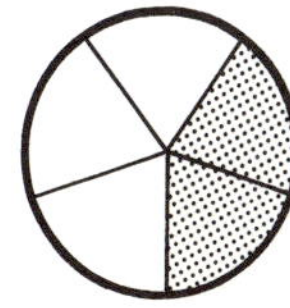

Would you be happy with this arrangement? We hope not.

Take the pretest and let's find out the extent of your knowledge and skill with fractions.

Add:

$\frac{3}{4} + \frac{3}{8} =$ ______ $1\frac{2}{3} + \frac{3}{5} =$ ______ $2\frac{3}{8} + 1\frac{1}{2} =$ ______

Subtract:

$\frac{3}{4} - \frac{3}{8} =$ ______ $1\frac{2}{3} - \frac{3}{5} =$ ______ $2\frac{3}{8} - 1\frac{1}{2} =$ ______

Multiply:

$\frac{3}{4} \times \frac{3}{8} =$ ______ $1\frac{2}{3} \times \frac{3}{5} =$ ______ $2\frac{3}{8} \times 1\frac{1}{2} =$ ______

Divide:

$\frac{3}{4} \div \frac{3}{8} =$ ______ $1\frac{2}{3} \div \frac{3}{5} =$ ______ $2\frac{3}{8} \div 1\frac{1}{2} =$ ______

Reduce to lowest terms:

$\frac{8}{32} =$ ______ $\frac{15}{39} =$ ______ $\frac{15}{48} =$ ______

Find the missing numerator:

$\frac{5}{6} = \frac{\quad}{24}$ $\frac{3}{7} = \frac{\quad}{35}$

Find the missing denominator:

$\frac{3}{5} = \frac{15}{\quad}$ $\frac{5}{8} = \frac{35}{\quad}$

THE MEANING OF A FRACTION

To "fracture" a leg is to break the leg bone into smaller pieces.

The word "fraction" means the pieces made when a whole unit is broken into smaller parts.

Below we have a picture of a dollar bill.

$1

If the dollar is cut in two, we will get two half-dollars.

The numerical symbol for a half is 1/2.

half dollar + half dollar = 1 dollar
1/2 + 1/2 = 2/2 = 1

If we cut the dollar into four equal pieces, we will get four quarter-dollars.

A quarter-dollar is usually called a "quarter."

A quarter is one-fourth the value of 1 dollar.

The numerical symbol for a quarter is 1/4.

quarter + quarter + quarter + quarter = 1 dollar
1/4 + 1/4 + 1/4 + 1/4 = 4/4 = 1

If you are given 4 quarters and one of them is removed, you would have only 3 quarters left.

The fraction that is being removed is ....1/4.
The faction that remains is ....3/4.

3/4 + 1/4 = 1

WHAT IS THE FRACTION TRYING TO TELL US?

In the fraction 3/4, the 4 tells us how many equal parts are in the whole unit.

The number 3 tells us how many of these parts we are concerned about.

When we write the fraction 3/4,

the number written above the line is called the NUMERATOR;

the number written below the line is called the DENOMINATOR.

1	2	3	4	5
6	7	8	9	10

The cake above has been cut into 10 equal pieces.

Each piece is what fraction of the whole unit? 1/10

What fraction of the cake is shaded? 6/10

What fraction of the cake is not shaded? 4/10

6/10 + 4/10 = 10/10 = 1

The DENOMINATOR of the fraction 6/10 is there
to tell us that the whole unit
was cut up into 10 equal pieces.

The NUMERATOR of the fraction 6/10 is there
to tell us that we are concerned
with only 6 of these pieces.

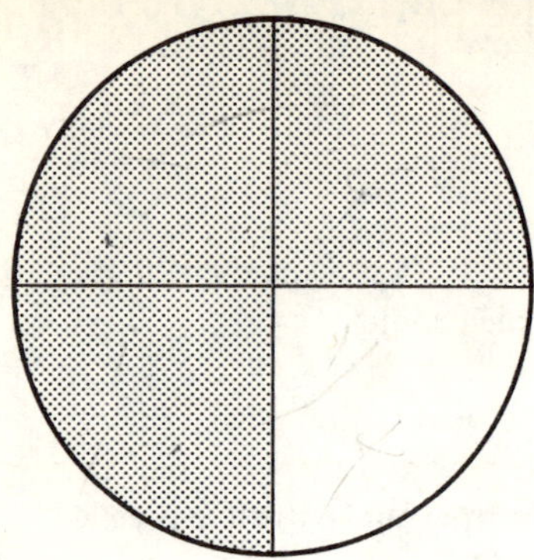

The circle has been cut up into how many equal pieces? ______
Each piece is what fraction of the whole circle? ______
What fraction of the circle is shaded? ______
What fraction of the circle is not shaded? ______
3/4 + 1/4 = ______ ; 4/4 = ______
In the fraction 3/4,
the numerator is ______
the denominator is ______
In the fraction 3/4, the denominator tells us that the whole unit is cut up into how many equal parts? ______
The denominator tells us that each piece is what fraction of the whole unit? ______
In the fraction 3/4, the numerator tells us that we are concerned with how many parts of the whole unit? ______

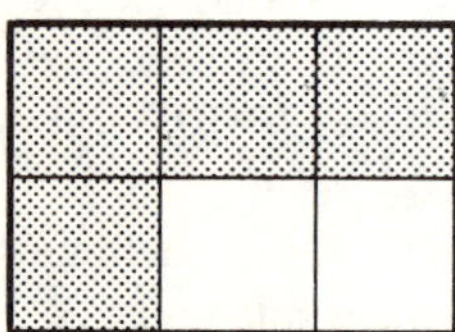

The cake above has been cut up into how many equal pieces? ______
Each piece is what fraction of the whole cake? ______
What fraction of the cake is shaded? ______
What fraction of the cake is not shaded? ______
4/6 + 2/6 = ______ ; 6/6 = ______
In the fraction 4/6,
the numerator is ______
the denominator is ______
In the fraction 4/6, the denominator tells us that the whole cake was cut up into how many equal parts? ______
The denominator tells us that each piece is what fraction of the whole cake? ______
In the fraction 4/6, the numerator tells us that we are concerned with how many parts of the whole unit? ______

This tray of candy contains how many pieces? ______

Each piece is what fraction of the whole tray? ______

What fraction of the tray is shaded? ______

What fraction of the tray is not shaded? ______
7/12 + 5/12 = ______ ; 12/12 = ______
In the fraction 7/12,
the numerator is ______
the denominator is ______
In the fraction 7/12, the denominator tells us that the whole unit is made up into how many equal pieces? ______
The denominator tells us that each piece is what fraction of the whole unit? ______
In the fraction 7/12, the numerator tells us that we are concerned with how many pieces of the whole tray? ______

3/5 is equal to ______ 1/5s
5/8 is equal to ______ 1/8s
2/3 is equal to ______ 1/3s
5/6 is equal to ______ 1/6s
7/10 is equal to ______ 1/10s

EQUIVALENT FRACTIONS

Below we have a picture of a dollar bill being equal to four quarters.

Each quarter is what part of 1 dollar? 1/4

It will take how many quarters to make a 1/2 dollar? 2

You can see that 2/4 is the equal of 1/2.

Another word for equal is "equivalent."

We say that 2/4 is the equivalent of 1/2.

The fractions 2/4 and 1/2 are called EQUIVALENT FRACTIONS because the two fractions represent the same value.

Here is a picture of a cake that has been cut into 8 pieces.

1/4	2/4	3/4	4/4
1/8	1/8	1/8	1/8
1/8	1/8	1/8	1/8

Each piece is what part of the whole cake? 1/8

If you were to receive 1/4 of the cake, how many 1/8s would you get? 2

That is, 2/8 = 1/4

The fractions 2/8 and 1/4 are called EQUIVALENT FRACTIONS.

If you were to receive 1/2 of the cake, how many 1/8s would you get? 4

That is, 4/8 = 1/2
The fractions 4/8 and 1/2 are also called EQUIVALENT FRACTIONS.

EACH FRACTION CAN HAVE MANY NAMES

Below we have an expanded 1-inch section of a ruler.

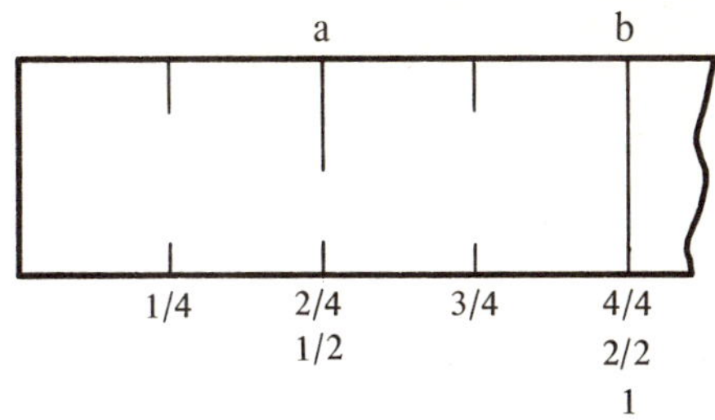

It is divided into four equal divisions; therefore, each division represents 1/4 of an inch.
Position a may be identified by two names,
2/4 or 1/2

Position b may be identified by three names,
4/4, 2/2 or 1

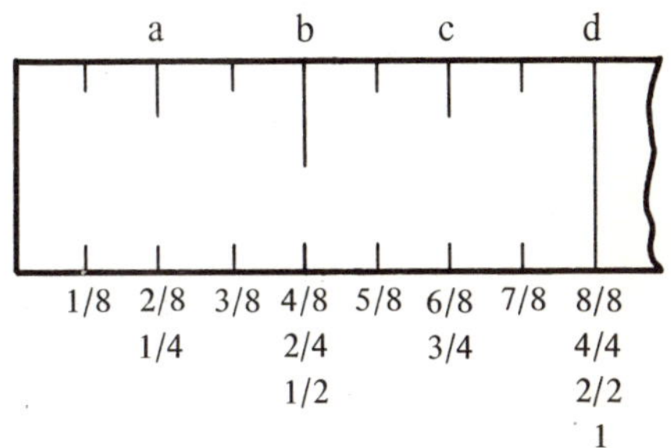

This time the one inch is divided into 8 equal divisions.
Each division represents 1/8 of an inch.

Position a may be identified as,
2/8 or 1/4
Position b may be identified as,
4/8, 2/4 or 1/2
Position c may be identified as,
6/8 or 3/4
Position d may be identified as,
8/8, 4/4, 2/2 or 1

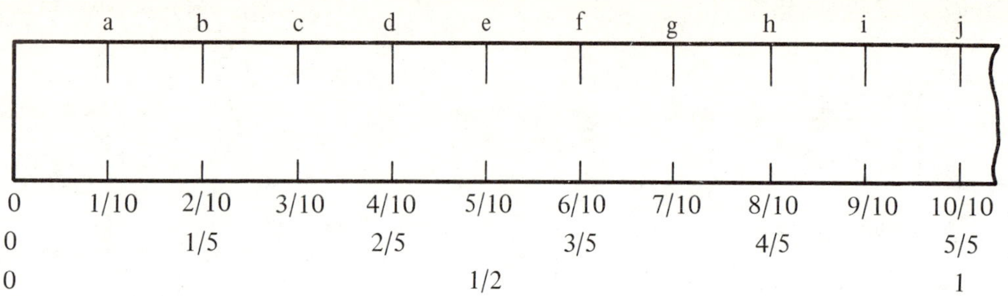

The above is an expanded 1-inch section of a ruler.

It has been divided into how many equal divisions? ____________

Each division represents what part of an inch? ____________

Position a may be identified as the fraction _____

Position b may be identified as the fraction _____ or _____

Position c may be identified as the fraction _____

Position d may be identified as the fraction _____ or _____

Position e may be identified as the fraction _____ or _____

Position f may be identified as the fraction _____ or _____

Position g may be identified as the fraction _____

Position h may be identified as the fraction _____ or _____

Position i may be identified as the fraction _____

Position j may be identified as the fraction _____ or _____ or as _____

The fractions 2/10 and 1/5 are called ____________ fractions
because they both represent the ____________.

The fractions 4/10 and ______ are also called equivalent fractions
and so are 6/10 and ______.
8/10 and ______.
10/10 and ______.

If we add 2/10 and 8/10, we should get ________.

If we add 2/5 and 3/5, we should get ________.

The pie at the right has been cut into how many slices? ________

Each slice will be what part of the whole pie? ________

Write the fraction that can be used to represent the shaded areas. ________

Write the fraction that can be used to represent
the unshaded areas. ________

The sum of the shaded areas and the unshaded areas (2/6 + 4/6)
must equal ________.

And 6/6 is equal to ________.

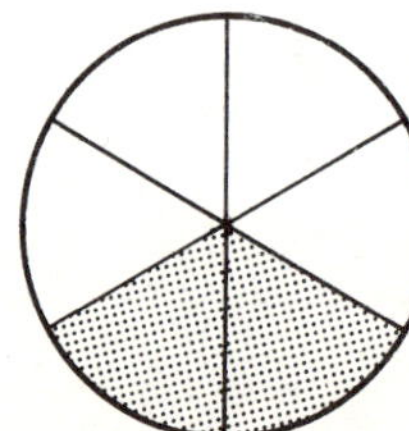

CHANGING THE FORM OF A FRACTION

When we studied the multiplication operation, we discovered that

any number multiplied by 1 equals the number.

2 x 1 = 2
15 x 1 = 15
575 x 1 = 575 and so on

Does it also hold true for fractions? Of course it does, because a fraction is a number.

1/2 x 1 = 1/2
7/8 x 1 = 7/8
15/3 x 1 = 15/3

When we studied the division operation, we discovered that
any number divided by itself equals 1.

2/2 = 1
15/15 = 1
250/250 = 1

Since 1/2 x 1 = 1/2

then 1/2 x 2/2 = 1/2
1/2 x 15/15 = 1/2
1/2 x 250/250 = 1/2 and so on

When you multiply 1/2 by 2/2, you are really multiplying 1/2 by 1, and the value of the fraction remains unchanged.

However, if we actually multiply both numerator and denominator by 2 we will CHANGE ITS FORM to read:

1/2 x 2/2 = 2/4

We can say the 1/2 and 2/4 are EQUIVALENT FRACTIONS because, even though they contain different numbers, they both represent the same value.

EQUIVALENT FRACTIONS IN HIGHER TERMS

The numbers involved in a fraction are often called "terms."

Higher terms simply means "larger numbers."

1/2 x 4/4 = 4/8 4/8 = 1/2

1/2 x 8/8 = 8/16 8/16 = 1/2

To change any fraction to an "equivalent fraction in higher terms," MULTIPLY both numerator and denominator by the same number.

OTHER EXAMPLES:

3/4 x 2/2 = 6/8 6/8 = 3/4
3/5 x 5/5 = 15/25 15/25 = 3/5
4/7 x 8/8 = 32/56 32/56 = 4/7

GUESSING GAMES

We'd like to change the form of 3/8 to an equivalent fraction with a denominator of 32.

3/8 = ?/32

Question How does one find the numerator of the equivalent fraction?

Answer If we find out what number 8 was multiplied by to get 32, then all we'd have to do is multiply the old numerator by the same number to get the new numerator.
In the problem above, the number is 4.

Therefore 3/8 = 12/32

BY THE WAY we could have gotten the same number (4) by dividing 32 by 8, right?

EXAMPLE: 3/5 = 18/?

18 divided by 3 = 6

Therefore 3/5 = 18/30

TIME LIMIT—3 Minutes

FRACTIONS

Find the missing term:

$\frac{1}{2} = \frac{2}{?}$	$\frac{1}{2} = \frac{8}{?}$	$\frac{1}{4} = \frac{4}{?}$	$\frac{1}{4} = \frac{3}{?}$	$\frac{3}{4} = \frac{?}{12}$
$\frac{3}{10} = \frac{?}{30}$	$\frac{1}{4} = \frac{?}{60}$	$\frac{3}{20} = \frac{?}{100}$	$\frac{2}{3} = \frac{?}{12}$	$\frac{1}{2} = \frac{6}{?}$
$\frac{1}{9} = \frac{?}{72}$	$\frac{5}{6} = \frac{?}{18}$	$\frac{3}{8} = \frac{?}{32}$	$\frac{12}{32} = \frac{24}{?}$	$\frac{1}{3} = \frac{?}{27}$
$\frac{4}{5} = \frac{20}{?}$	$\frac{3}{16} = \frac{?}{32}$	$\frac{3}{16} = \frac{15}{?}$	$\frac{1}{4} = \frac{?}{64}$	$\frac{4}{5} = \frac{16}{?}$
$\frac{8}{32} = \frac{?}{64}$	$\frac{12}{32} = \frac{36}{?}$	$\frac{1}{6} = \frac{?}{24}$	$\frac{3}{4} = \frac{24}{?}$	$\frac{3}{10} = \frac{?}{50}$
$\frac{12}{20} = \frac{60}{?}$	$\frac{1}{6} = \frac{?}{72}$	$\frac{3}{5} = \frac{?}{40}$	$\frac{11}{16} = \frac{55}{?}$	$\frac{1}{7} = \frac{?}{56}$

SCORING

1 error = excellent
2 errors = good

REDUCING THE FORM OF A FRACTION

When we studied the division operation we discovered that

any number divided by 1 equals the number.

$$2/1 = 2$$
$$30/1 = 30$$
$$400/1 = 400$$

Does this also hold true for fractions?
Of course it does; fractions are numbers.

$$\frac{10}{20} \div 1 = 10/20$$

$$\frac{12}{16} \div 1 = 12/16$$

$$\frac{15}{25} \div 1 = 15/25 \quad \text{and so on}$$

We also know that the number 1 may take many forms:

$$2/2 = 1$$
$$3/3 = 1$$
$$4/4 = 1 \quad \text{and so on}$$

Then
$$10/20 \div 2/2 = 10/20$$
$$12/16 \div 4/4 = 12/16$$
$$15/25 \div 5/5 = 15/25$$

When we divide 10/20 by 2/2 we are actually dividing 10/20 by 1, and the value of the fraction remains unchanged.

However, if we actually divide both numerator and denominator by 2, we will get:

$$\frac{10}{20} \div \frac{2}{2} = \frac{5}{10}$$

We can say that 10/20 and 5/10 are EQUIVALENT FRACTIONS.

OTHER EXAMPLES:

$$\frac{12}{16} \div \frac{4}{4} = \frac{3}{4} \qquad \frac{12}{16} = \frac{3}{4} \qquad \frac{15}{25} \div \frac{5}{5} = \frac{3}{5} \qquad \frac{15}{25} = \frac{3}{5}$$

REDUCING THE FRACTION TO ITS LOWEST TERMS

Reducing a fraction to "lower terms" is one thing, and reducing a fraction to its "lowest terms" is another.

EXAMPLE: Divide both numerator and denominator of the fraction 24/36 by 2, and you get in "lower terms" 12/18

However, to reduce the fraction 24/36 to its "lowest terms" we must find the LARGEST number that can be divided evenly into both numerator and denominator.

THINK

Is 2 the largest? NO, because 12/18 can still be reduced.
Is 3 the largest? NO, because 8/12 can still be reduced.
Is 4 the largest? NO, because 6/9 can still be reduced.
Is 6 the largest? NO, because 4/6 can still be reduced.
Is 12 the largest? YES, because 2/3 cannot be reduced any further.

2/3 is 24/36 in its LOWEST TERMS.

OTHER EXAMPLES:

$$\frac{18}{45} \div \frac{9}{9} = \frac{2}{5}$$

$$\frac{42}{77} \div \frac{7}{7} = \frac{6}{11}$$

GUESSING GAMES

$$\frac{12}{20} = \frac{?}{5}$$

The denominator of the first fraction is 4 times the denominator of the second fraction. The numerator of the first fraction must also be 4 times the numerator of the second.
The missing numerator is 3.

Supply the missing term:

$$\frac{27}{36} = \frac{3}{?}$$

The missing denominator is 4.

TIME LIMIT–5 Minutes

FRACTIONS

Reduce to lowest terms:

$\frac{12}{64} =$	$\frac{12}{32} =$	$\frac{9}{36} =$	$\frac{20}{60} =$	$\frac{8}{12} =$
$\frac{4}{40} =$	$\frac{40}{100} =$	$\frac{40}{48} =$	$\frac{8}{24} =$	$\frac{7}{35} =$
$\frac{15}{20} =$	$\frac{18}{24} =$	$\frac{28}{49} =$	$\frac{8}{20} =$	$\frac{10}{24} =$
$\frac{60}{72} =$	$\frac{9}{54} =$	$\frac{20}{24} =$	$\frac{18}{60} =$	$\frac{60}{64} =$
$\frac{20}{32} =$	$\frac{21}{48} =$	$\frac{42}{72} =$	$\frac{11}{110} =$	$\frac{20}{50} =$

SCORING

1 error = excellent
2 errors = good

PROPER AND IMPROPER FRACTIONS

 $= \frac{2}{3}$  $= \frac{3}{4}$

Fractions such as $\frac{2}{3}$ and $\frac{3}{4}$

represent a part of a whole unit, and because they truly represent a part of a whole unit they are called PROPER FRACTIONS.

 $= \frac{3}{3}$  $= \frac{4}{4}$

Fractions such as $\frac{3}{3}$ and $\frac{4}{4}$

represent a whole unit, and because they do not truly represent a part of a whole unit, they are called IMPROPER FRACTIONS.

 +  $= \frac{4}{3}$  +  $= \frac{7}{4}$

Fractions such as $\frac{4}{3}$ and $\frac{7}{4}$

represent a value larger than a whole unit, and because they too do not truly represent a part of a whole unit, they are also called IMPROPER FRACTIONS.

SUMMARY: When the numerator is smaller than the denominator, the fraction is called a PROPER FRACTION.
When the numerator is equal to or larger than the denominator, the fraction is called an IMPROPER FRACTION.

MIXED NUMBERS

 + 

Each figure has been divided into quarters.
2 whole units = 8 quarters

There are seven (7) shaded areas $= \frac{7}{4}$

$$\frac{7}{4} = \frac{4}{4} + \frac{3}{4}$$

since 4/4 is equal to 1,

then $\frac{7}{4} = 1 + \frac{3}{4}$

It is customary to write $1 + \frac{3}{4}$ as $1\frac{3}{4}$

$\frac{7}{4}$ is called an IMPROPER FRACTION.

$1\frac{3}{4}$ is called a MIXED NUMBER.

CHANGING IMPROPER FRACTIONS TO MIXED NUMBERS

When the numerator (dividend) is evenly divided by the denominator (divisor), we will get a whole number for an answer.

EXAMPLES: $\frac{4}{4} = 1$, $\frac{9}{3} = 3$, $\frac{25}{5} = 5$ and so on

When the numerator is not evenly divided by the denominator, we will get a remainder.

EXAMPLES: $\frac{5}{2} = 2\frac{1}{2}$, $\frac{11}{3} = 3\frac{2}{3}$, $\frac{19}{4} = 4\frac{3}{4}$

The remainder is written as a proper fraction with the remainder as the numerator and the denominator identically the same as in the improper fraction.

NOTE: The fractional part of a mixed number should always be reduced to its lowest terms.

FRACTIONS

Change the improper fractions to mixed numbers; reduce fractions to lowest terms.

$\frac{20}{7} =$	$\frac{19}{2} =$	$\frac{28}{3} =$	$\frac{55}{15} =$	$\frac{13}{5} =$
$\frac{38}{10} =$	$\frac{43}{4} =$	$\frac{64}{16} =$	$\frac{18}{3} =$	$\frac{51}{16} =$
$\frac{105}{32} =$	$\frac{43}{3} =$	$\frac{18}{4} =$	$\frac{73}{32} =$	$\frac{87}{16} =$
$\frac{52}{5} =$	$\frac{15}{6} =$	$\frac{64}{8} =$	$\frac{67}{15} =$	$\frac{37}{12} =$
$\frac{35}{8} =$	$\frac{36}{7} =$	$\frac{41}{12} =$	$\frac{49}{8} =$	$\frac{40}{7} =$
$\frac{48}{5} =$	$\frac{47}{11} =$	$\frac{73}{11} =$	$\frac{28}{9} =$	$\frac{100}{3} =$

SCORING

1 error = excellent
2 errors = good

CHANGING WHOLE NUMBERS TO IMPROPER FRACTIONS

How many quarters will make 2 dollars?	8
How many 1/4s in the number 2?	8/4
How many dimes will make 3 dollars?	30
How many 1/10s in the number 3?	30/10

Converting whole numbers to improper fractions is no harder than changing money from one unit to another.

PROCEDURE:

1. Select the desired denominator.
2. To get the numerator, multiply the whole number by the desired denominator.

EXAMPLES:

Change the whole number 5 to an improper fraction having a denominator of 3.

5 = 5 x 3/3 = 15/3

Change the whole number 8 to an improper fraction having a denominator of 5.

8 = 8 x 5/5 = 40/5

How many eighths of an inch are there in 4 inches?

4 = 4 x 8/8 = (32 eighths)

How many quarts in 6 gallons?

6 = 6 x 4/4 = (24 quarts)

How many hours in 3 days?

3 = 3 x 24/24 = (72 hours)

CHANGING MIXED NUMBERS TO IMPROPER FRACTIONS

How many halves will make 3 1/2 dollars?	7
How many 1/2s in the mixed number 3 1/2?	7/2
How many quarters will make 2 1/4 dollars?	9
How many 1/4s in the mixed number 2 1/4?	9/4

Converting mixed numbers to improper fractions is no harder than changing money from one unit to another.

PROCEDURE: Convert 3 3/4 to an improper fraction.

1. Multiply the whole number by the denominator of the fractional part of the mixed number (4 x 3 = 12). $3\frac{3}{4}$
2. Add the product (12) to the numerator of the fractional member of the mixed number (12 + 3 = 15). $3\frac{3}{4}$
3. The denominator of the improper fraction will be the denominator of the fractional member of the mixed number (4). $3\frac{3}{4} = \frac{15}{4}$

EXAMPLES:

Change the following mixed numbers to improper fractions:

$$4\frac{2}{3} = \left(\frac{3 \times 4}{3}\right) + \left(\frac{2}{3}\right) = \frac{14}{3}$$

$$6\frac{3}{5} = \left(\frac{5 \times 6}{5}\right) + \left(\frac{3}{5}\right) = \frac{33}{5}$$

$$12\frac{4}{7} = \left(\frac{7 \times 12}{7}\right) + \left(\frac{4}{7}\right) = \frac{88}{7}$$

TIME LIMIT–4 Minutes

FRACTIONS

Change the following whole numbers to improper fractions with the following desired denominators:

$3 = \frac{\quad}{8}$	$2 = \frac{\quad}{5}$	$4 = \frac{\quad}{6}$	$4 = \frac{\quad}{9}$	$7 = \frac{\quad}{7}$
$8 = \frac{\quad}{12}$	$5 = \frac{\quad}{15}$	$6 = \frac{\quad}{20}$	$6 = \frac{\quad}{11}$	$2 = \frac{\quad}{13}$
$12 = \frac{\quad}{8}$	$13 = \frac{\quad}{7}$	$15 = \frac{\quad}{9}$	$20 = \frac{\quad}{6}$	$25 = \frac{\quad}{5}$

Change the following mixed numbers to improper fractions:

4 4/5 =	4 5/12 =	3 5/18 =	12 3/4 =
7 3/16 =	3 13/16 =	4 3/40 =	28 1/2 =
8 3/7 =	9 1/3 =	11 2/7 =	33 1/3 =
10 1/6 =	22 1/6 =	13 3/8 =	7 2/13 =
5 4/9 =	2 13/20 =	2 3/50 =	14 3/16 =

SCORING

1 error = excellent
2 errors = good

ADDITION OF LIKE FRACTIONS

The fractions 3/4 and 1/4 have the same denominator.

Fractions that have the same denominator are said to have a COMMON denominator.

Two or more fractions having the same denominator are called LIKE FRACTIONS.

1 quarter and 2 quarters can be combined to give us a total of

1 quarter
2 quarters
3 quarters

3 quarters and 2 dimes cannot be combined until they are changed to

75 cents
20 cents
95 cents

RULE: The addition operation obeys one and only one rule:
You can only add LIKE units.

Take whole numbers, for example;
You can add only:
ones to ones
tens to tens
hundreds to hundreds and so on

and so with fractions,
you can add only fractions whose denominators are alike (LIKE FRACTIONS).

For example:
3ds to 3ds
4ths to 4ths
8ths to 8ths and so on.

VISUAL ADDITION

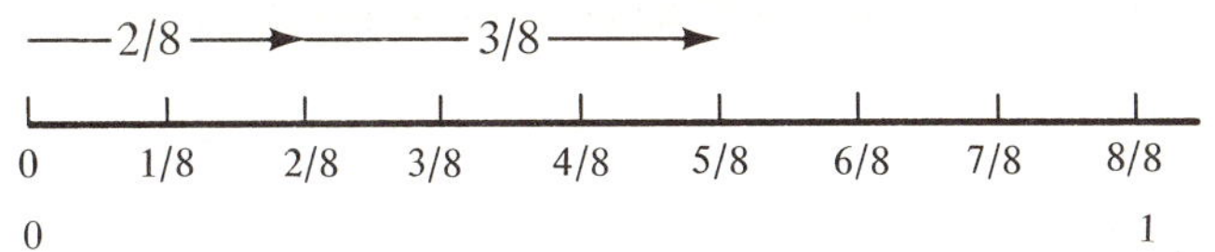

RULE: The sum of LIKE fractions can be found by adding their numerators.

EXAMPLE 1

2/7 + 3/7 = 5/7

1/6 + 4/6 = 5/6

EXAMPLE 2

3/5 + 2/5 = 5/5 = 1

3/8 + 7/8 = 10/8 = 1 2/8 = 1 1/4

5/6 + 5/6 = 10/6 = 1 4/6 = 1 2/3

RULE: When the sum of two or more fractions becomes an IMPROPER FRACTION, it should always be changed to a WHOLE NUMBER—or a MIXED NUMBER—and the fractional part of the mixed number reduced to LOWEST TERMS.

EXAMPLE 3

Can we use the rules above for the addition of three or more fractions?
YES—for as many as you wish.

3/8 + 4/8 + 5/8 = 12/8 = 1 4/8 = 1 1/2

2/6 + 1/6 + 5/6 + 4/6 = 12/6 = 2

ADDITION OF FRACTIONS

Reduce answers to lowest terms.

$\frac{1}{8}$ $\frac{3}{7}$ $\frac{3}{5}$ $\frac{5}{6}$ $\frac{4}{9}$ $\frac{7}{10}$

$\frac{5}{8}$ $\frac{4}{7}$ $\frac{4}{5}$ $\frac{5}{6}$ $\frac{7}{9}$ $\frac{8}{10}$

$\frac{3}{8}$ $\frac{5}{7}$ $\frac{3}{5}$ $\frac{3}{6}$ $\frac{4}{9}$ $\frac{9}{10}$

$\frac{7}{8}$ $\frac{3}{7}$ $\frac{3}{5}$ $\frac{3}{6}$ $\frac{5}{9}$ $\frac{6}{10}$

$\frac{5}{8}$ $\frac{3}{7}$ $\frac{4}{5}$ $\frac{1}{6}$ $\frac{3}{9}$ $\frac{3}{10}$

$\frac{7}{8}$ $\frac{4}{7}$ $\frac{2}{5}$ $\frac{2}{6}$ $\frac{5}{9}$ $\frac{9}{10}$

$\frac{3}{8}$ $\frac{4}{9}$ $\frac{3}{5}$ $\frac{5}{6}$ $\frac{4}{7}$ $\frac{7}{10}$

$\frac{5}{8}$ $\frac{7}{9}$ $\frac{1}{5}$ $\frac{3}{6}$ $\frac{6}{7}$ $\frac{9}{10}$

$\frac{7}{8}$ $\frac{3}{9}$ $\frac{4}{5}$ $\frac{4}{6}$ $\frac{3}{7}$ $\frac{3}{10}$

SCORING

1 error = excellent
2 errors = good

ADDITION OF LIKE FRACTIONS (cont.)

Can a mixed number be added to a fraction? YES—providing the fractional members have LIKE denominators.

EXAMPLE 4

```
2 2/9
  4/9
-----
```

The mixed number 2 2/9 means which is to be added to

```
2 + 2/9
    4/9
-------
```

Addition permits us to add the whole numbers and fractions in any order. In this example it is convenient to:

1. Add the fractions first. = 6/9
2. Combine the answer with the whole number. = 2 6/9
3. Reduce the answer to its lowest terms. = 2 2/3

Can a mixed number be added to another mixed number? YES—providing the fractional members have LIKE denominators.

EXAMPLE 5

```
2 3/8
1 3/8
-----
3 6/8 = 3 3/4
```

```
3 4/8
1 4/8
-----
4 8/8 = 5
```

RULE:

1. Add the fractions.
2. Add the whole numbers.
3. Combine the results into a mixed number.
4. Reduce to lowest terms.

EXAMPLE 6

```
5 5/9
3 7/9
-----
9 1/3
```

1. Add the whole numbers. = 8
2. Add the fractions. = 12/9
3. Convert to a mixed number. = 1 3/9
4. The 1 3/9 combined with the sum of whole numbers makes = 9 3/9
5. Reduce the answer to its lowest terms. = 9 1/3

Can a combination of fractions, mixed numbers, and whole numbers be added together? YES—providing the fractional members have LIKE denominators.

EXAMPLE 7

```
2 3/4
  2/4
3
  1/4
-----
6 1/2
```

1. Add the whole numbers. = 5
2. Add the fractions. = 6/4
3. Convert to a mixed number. = 1 2/4
4. The 1 2/4 combined with the sum of whole numbers makes = 6 2/4
5. Reduce the answer to lowest terms. = 6 1/2

In some areas the horizontal method is preferred over the vertical method.

EXAMPLE 8 3 3/8 + 2 1/8 + 1 5/8 = 7 1/8

1. Convert each mixed number to an improper fraction and add.

$$\frac{27}{8} + \frac{17}{8} + \frac{13}{8} = \frac{57}{8}$$

2. Convert to a mixed number. = 7 1/8

ADDITION OF FRACTIONS

Reduce answers to lowest terms.

1 2/9 5/9	4 1/6 3/6	3 1/7 4/7	5 5/16 9/16	2 1/8 5/8
2 2/5 3 1/5	4 3/7 2 2/7	5 1/8 3 3/8	2 2/9 6 4/9	6 3/10 5 2/10
3 2/3 2 1/3	4 4/7 2 3/7	5 3/8 1 5/8	2 1/5 2 4/5	4 4/10 5 6/10
4 2/3 3 2/3	3 7/10 5 6/10	6 5/8 4 5/8	5 5/6 3 3/6	1 9/10 3 6/10
2 1/8 3/8 5 5/8	3 3/16 7/16 8 7/16	5 3/5 2 4/5 3/5	6 5/12 3/12 4 7/12 1/12	5/6 3 5/6 1/6 4 2/6

SCORING

1 error = excellent
2 errors = good

ADDITION OF TWO UNLIKE FRACTIONS

REMEMBER: Addition obeys one and only one rule: You can add only LIKE units.

LIKE units in fractions are those that have identical denominators.

Well then, how does one add UNLIKE FRACTIONS?

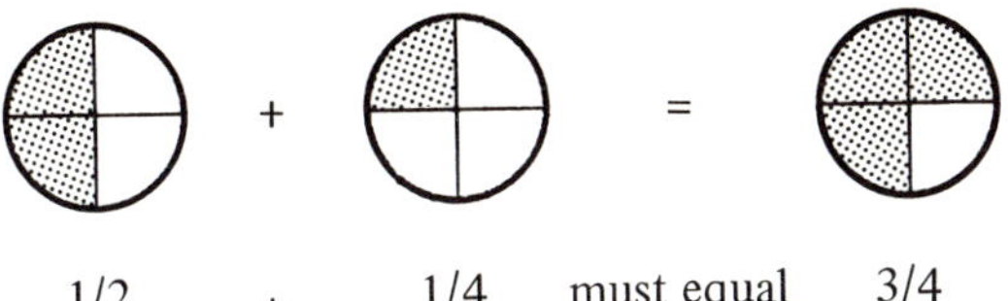

1/2 + 1/4 must equal 3/4

—1/2—→—1/4→

0 1/4 2/4 3/4 4/4

0 1

The fraction 1/2 cannot be added directly to 1/4. Why not? Because they do not have the same denominators.

What must be done? Change them to a common unit.
We must either change the 1/4 to 1/2s,
which cannot be done—or
change the 1/2 to 1/4s, which can be done.

To change 1/2 to an equivalent fraction having a denominator of 4 we must multiply the fraction by 2/2, which has a value of 1, and you know, that will not change the value of the fraction.

$$\frac{1}{2} \times \frac{2}{2} = \frac{1 \times 2}{2 \times 2} = \frac{2}{4}$$

Now, we can add $\frac{2}{4} + \frac{1}{4}$ to get $\frac{3}{4}$

BY THE WAY, did you know that $\frac{2}{4} + \frac{1}{4}$

may be written as $\frac{2+1}{4}$

FINDING A COMMON DENOMINATOR

Several methods are available to us—let's explore them.

METHOD 1 $\frac{1}{2} + \frac{1}{3} =$

a. Multiply the denominators (2 and 3) to get the COMMON DENOMINATOR = 6

b. To change 1/2 to a fraction having a denominator of 6, multiply both members by 3.

c. To change 1/3 to a fraction having a denominator of 6, multiply both members by 2.

$$\left[\frac{1}{2} \times \frac{3}{3}\right] + \left[\frac{1}{3} \times \frac{2}{2}\right] =$$

$$\frac{3}{6} + \frac{2}{6} = \frac{3+2}{6} = \frac{5}{6}$$

ANOTHER EXAMPLE $\frac{2}{3} + \frac{3}{4} =$

a. The COMMON denominator of 3 and 4 is 12

b. To change 2/3 to a fraction having a denominator of 12, multiply both members by 4.

c. To change 3/4 to a fraction having a denominator of 12, multiply both members by 3.

$$\left[\frac{2}{3} \times \frac{4}{4}\right] + \left[\frac{3}{4} \times \frac{3}{3}\right] =$$

$$\frac{8}{12} + \frac{9}{12} = \frac{8+9}{12} = \frac{17}{12}$$

d. Convert to a mixed number. 1 5/12

TIME LIMIT—10 Minutes

ADDITION OF FRACTIONS

Reduce answers to lowest terms.

1/2 + 1/3 =	1/2 + 1/4 =	1/2 + 1/5 =	1/2 + 1/6 =
1/2 + 1/7 =	1/2 + 1/10 =	1/2 + 1/8 =	1/2 + 1/9 =
1/3 + 1/4 =	1/3 + 1/5 =	1/3 + 1/6 =	1/3 + 1/7 =
1/4 + 1/5 =	1/4 + 1/6 =	1/4 + 1/7 =	1/4 + 1/8 =
3/4 + 2/5 =	2/3 + 3/5 =	3/4 + 5/6 =	4/5 + 5/6 =
2/3 + 5/6 =	1/2 + 3/7 =	3/4 + 3/7 =	2/3 + 3/7 =
3/5 + 3/7 =	5/6 + 3/7 =	1/2 + 5/8 =	2/3 + 5/8 =
3/4 + 5/8 =	4/5 + 5/8 =	5/6 + 5/8 =	2/3 + 4/9 =
3/4 + 4/9 =	4/5 + 4/9 =	5/6 + 4/9 =	6/7 + 4/9 =

SCORING

1 error = excellent
2 errors = good

NEW NAMES

On the pages that follow we shall describe certain new procedures using some new names, and we think it's best we take some time out now and introduce them to you.

PRIME NUMBERS

A PRIME NUMBER is a number that cannot be divided evenly by any other number except itself and the number 1.

1, 2, 3, 5, 7, 11, 13, 17, 19, 23, and so on

are example of prime numbers.

COMPOSITE NUMBERS are what we call all other numbers.

A COMPOSITE NUMBER is a number that is considered to be the product of two or more prime numbers.

EXAMPLES:

4–the product of the prime numbers 2 x 2
6–the product of the prime numbers 2 x 3
8–the product of the prime numbers 2 x 2 x 2
9–the product of the prime numbers 3 x 3
10–the product of the prime numbers 2 x 5
12–the product of the prime numbers 2 x 2 x 3
14–the product of the prime numbers 2 x 7
15–the product of the prime numbers 3 x 5

MULTIPLES

The multiple of a number is the product of that number and any other number.

EXAMPLES:

4, 6, 8, 10, 12, 14, 16 are multiples of 2.
6, 9, 12, 15, 18, 21, 24 are multiples of 3.
8, 12, 16, 20, 24, 28 are multiples of 4.
10, 15, 20, 25, 30, 35 are multiples of 5.

COMMON MULTIPLES

The multiples of 3 are
6, 9, 12, 15, 18, 21, 24, 27, 30 and so on.

The multiples of 4 are
8, 12, 16, 20, 24, 28, 32, 36, 40 and so on.

The underscored numbers, 12 and 24, are COMMON MULTIPLES of the numbers 3 and 4.

PRIME FACTORS OF A NUMBER

EXAMPLE: Find the factors of 24 which are prime numbers.

PROCEDURE:

1. Start with the smallest prime factor other than 1. — 2 /24 → 12
2. Now factor 12 with its smallest prime factor. — 2 /12 → 6
3. Factor 6 with its smallest prime factor. — 2 / 6 → 3
4. 3 cannot be factored because it is a prime number.
5. The prime factors of 24 are 2 x 2 x 2 x 3
6. Check by multiplying the prime factors = 24

EXAMPLE: Find the prime factors of 28.

2 /28 → 14

2 /14 → 7

The prime factors of 28 are 2 x 2 x 7 = 28

TIME LIMIT—None

FRACTIONS

Circle all PRIME numbers:

2	3	4	5	6	7	8
9	10	11	12	13	14	15
16	17	18	19	20	21	22
23	24	25	26	27	28	29
30	31	32	33	34	35	36
37	38	39	40	41	42	43
44	45	46	47	48	49	50

What are the prime factors of each of the following? Verify by multiplying these factors.

16 ______________________
18 ______________________
20 ______________________
21 ______________________
22 ______________________
25 ______________________
26 ______________________
27 ______________________
30 ______________________

Give 9 MULTIPLES (in order) for each of the following numbers:

5 ______________________
7 ______________________
9 ______________________
12 ______________________

Give the numbers for which each of the following is a common MULTIPLE:

12 ______________________
15 ______________________
18 ______________________
20 ______________________
21 ______________________
24 ______________________

Find the factors that are PRIME numbers. Use the procedure shown in the lesson.

/ 24   / 56   / 36   / 48   / 81

ADDITION OF TWO UNLIKE FRACTIONS (cont.)

METHOD 1–Review

$$\frac{3}{4} + \frac{3}{8} =$$

$$\left[\frac{3}{4} \times \frac{8}{8}\right] + \left[\frac{3}{8} \times \frac{4}{4}\right] = \frac{24}{32} + \frac{12}{32} = \frac{36}{32}$$

Convert to a mixed number. 1 4/32
Reduce to lowest terms. 1 1/8

Was 32 the lowest common denominator? NO
The LOWEST common denominator was 8

The LOWEST COMMON DENOMINATOR is the smallest denominator that is a multiple of the other denominator (s).

How are we going to get the LOWEST common denominator (LCD) without going through all of the work above?

METHOD 2–Inspection Method

$$\frac{3}{4} + \frac{3}{8} =$$

Procedure:

1. Select the largest denominator. 8
2. Is it a multiple of the other denominator? YES
3. Then, the LCD is 8

EXAMPLE:

$$\frac{3}{4} + \frac{5}{6} =$$

1. The largest denominator is 6
2. Is it a multiple of 4? NO
3. DOUBLE it to get 12
4. Is it now a multiple of 4?. YES
5. Then 12 is the LCD.

$$\frac{3}{4} + \frac{5}{6} = \left[\frac{3}{4} \times \frac{3}{3}\right] + \left[\frac{5}{6} \times \frac{2}{2}\right] = \frac{19}{12} = 1\ 7/12$$

METHOD 3–For Larger Denominators

$$\frac{3}{24} + \frac{7}{18} =$$

A little too hard for Methods 1 or 2, don't you think?

PROCEDURE:

1. Place all denominators in line, as in short division, and separate them by commas. / 24, 18
2. Divide by the SMALLEST PRIME NUMBER (other than 1). 2 / 24, 18 — 12, 9
3. Divide again by the SMALLEST PRIME NUMBER. (The number 9, which is not a multiple of 2, is brought down unchanged.) 2 / 24, 18 — 2 / 12, 9 — 6, 9
4. Divide again by the SMALLEST PRIME NUMBER. 2 / 24, 18 — 2 / 12, 9 — 2 / 6, 9 — 3, 9
5. Divide again by the SMALLEST PRIME NUMBER. 2 / 24, 18 — 2 / 12, 9 — 2 / 6, 9 — 3 / 3, 9 — 1, 3
6. The division by the SMALLEST PRIME NUMBER continues–until the quotients in the last line are all 1. 2 / 24, 18 — 2 / 12, 9 — 2 / 6, 9 — 3 / 3, 9 — 3 / 1, 3 — 1, 1
7. Finally, the divisors (the smallest prime numbers) are multiplied together. Their product will be the LCD 72

$$\frac{3}{24} + \frac{7}{18} = \left[\frac{3}{24} \times \frac{3}{3}\right] + \left[\frac{7}{18} \times \frac{4}{4}\right] = \frac{37}{72}$$

8. The product is in its LOWEST terms.

TIME LIMIT–15 Minutes

ADDITION OF FRACTIONS

Use the procedure demonstrated in the lesson to find the LCD.
Reduce the answers to lowest terms.

2/3 + 5/12 =	3/4 + 7/16 =	5/6 + 7/10 =	5/9 + 5/12 =
5/8 + 3/20 =	5/8 + 5/12 =	5/6 + 7/8 =	3/4 + 7/12 =
5/14 + 7/15 =	9/12 + 15/48 =	3/8 + 5/32 =	3/15 + 7/20 =
5/7 + 5/28 =	7/12 + 3/10 =	5/8 + 9/10 =	9/12 + 13/18 =
7/18 + 5/14 =	9/30 + 11/24 =	5/24 + 13/16 =	11/18 + 3/4 =

SCORING

1 error = excellent
2 errors = good

ADDITION OF THREE OR MORE UNLIKE FRACTIONS

1/2 + 1/3 + 1/4 =

Find the LCD by Method 2 (Inspection Method).

1. The largest denominator is	4
2. Is it a multiple of the other denominators?	NO
3. DOUBLE it to get	8
4. Is it now a multiple of the others?	NO
5. TRIPLE it to get	12
6. Is it now a multiple of the others?	YES
7. Then the LCD is	12

Find the LCD by Method 3 (Prime Factor Method).

```
2 / 2, 3, 4
2 / 1, 3, 2
3 / 1, 3, 1
    1, 1, 1
```

The product of 2 x 2 x 3 = 12

The LCD is 12

SOLUTION: 1/2 + 1/3 + 1/4 =

$$\left[\frac{1}{2} \times \frac{6}{6}\right] + \left[\frac{1}{3} \times \frac{4}{4}\right] + \left[\frac{1}{4} \times \frac{3}{3}\right] =$$

$$\frac{6}{12} + \frac{4}{12} + \frac{3}{12} = \frac{13}{12} = 1\ 1/12$$

EXAMPLE: 5/6 + 7/12 + 2/9 =

It wouldn't always be possible to find the LCD by inspection. Your best tool is Method 3.

```
2 / 6, 12, 9
2 / 3,  6, 9
3 / 3,  3, 9
3 / 1,  1, 3
    1,  1, 1
```

The product of the prime numbers:
2 x 2 x 3 x 3 = 36

SOLUTION: $\frac{5}{6} + \frac{7}{12} + \frac{2}{9} =$

$$\frac{30}{36} + \frac{21}{36} + \frac{8}{36} = \frac{59}{36} = 1\ 23/36$$

CAUTION: If you were tempted to use numbers other than the prime numbers, you'll arrive at a common denominator—but not the LCD.

EXAMPLE:

```
6 / 6, 12, 9
2 / 1,  2, 9
9 / 1,  1, 9
    1,  1, 1
```

The product of the prime numbers:
6 x 2 x 9 = 108

Let's get into some deep water.

2/3 + 5/8 + 7/12 + 8/15 =

```
2 / 3, 8, 12, 15
2 / 3, 4,  6, 15
2 / 3, 2,  3, 15
3 / 3, 1,  3, 15
5 / 1, 1,  1,  5
    1, 1,  1,  1
```

The LCD is = 120

SOLUTION:

$$\frac{80}{120} + \frac{75}{120} + \frac{70}{120} + \frac{64}{120} = \frac{289}{120} = 2\ 49/120$$

TIME LIMIT–20 Minutes

ADDITION OF FRACTIONS

First, find the LCD by inspection or by the prime number method demonstrated in the lesson.

$\frac{1}{5} + \frac{3}{10} + \frac{4}{15} =$

$\frac{3}{6} + \frac{7}{12} + \frac{11}{24} =$

$\frac{1}{3} + \frac{5}{9} + \frac{7}{18} =$

$\frac{7}{10} + \frac{5}{12} + \frac{7}{15} =$

$\frac{3}{12} + \frac{2}{3} + \frac{5}{8} =$

$\frac{3}{7} + \frac{3}{4} + \frac{3}{28} =$

$\frac{5}{8} + \frac{5}{16} + \frac{3}{20} =$

$\frac{3}{8} + \frac{5}{9} + \frac{2}{3} =$

$\frac{11}{12} + \frac{7}{10} + \frac{3}{5} =$

$\frac{3}{8} + \frac{5}{16} + \frac{7}{12} =$

$\frac{3}{4} + \frac{7}{12} + \frac{7}{16} =$

$\frac{5}{6} + \frac{7}{8} + \frac{5}{12} =$

$\frac{14}{15} + \frac{5}{6} + \frac{4}{5} + \frac{3}{4} + \frac{11}{20} =$

$\frac{3}{8} + \frac{3}{16} + \frac{3}{32} + \frac{3}{64} =$

$\frac{11}{18} + \frac{5}{8} + \frac{7}{9} + \frac{13}{24} =$

$\frac{2}{3} + \frac{11}{18} + \frac{4}{5} + \frac{1}{2} =$

$\frac{3}{8} + \frac{11}{18} + \frac{13}{36} + \frac{25}{72} =$

$\frac{4}{5} + \frac{4}{9} + \frac{4}{15} + \frac{4}{30} =$

SCORING

1 error = excellent
2 errors = good

SUBTRACTION OF LIKE FRACTIONS

The subtraction operation resembles the addition operation in that all work is performed on the numerators.

SUBTRACTING A FRACTION FROM A FRACTION

$$\frac{7}{8} - \frac{1}{8} =$$

$$\frac{7}{8} - \frac{1}{8} = \frac{7-1}{8} = \frac{6}{8} \text{ (The difference)}$$

Reduce to lowest terms. $\frac{3}{4}$

PROOF: $\frac{6}{8} + \frac{1}{8} = \frac{7}{8}$

SUBTRACTING A MIXED NUMBER FROM A MIXED NUMBER

$$7\frac{3}{5} - 1\frac{1}{5} =$$

1. Subtract the fractions. $\frac{3}{5} - \frac{1}{5} = \frac{3-1}{5} = \frac{2}{5}$
2. Subtract the whole numbers. $7 - 1 = 6$
3. Combine the two. $6 + \frac{2}{5} = 6\frac{2}{5}$ (The difference)

PROOF: $6\frac{2}{5} + 1\frac{1}{5} = 7\frac{3}{5}$

SUBTRACTING A WHOLE NUMBER FROM A MIXED NUMBER

$$5\frac{5}{6} - 2 =$$

REMEMBER: 0 subtracted from any number does not change the value of the number.

1. $\frac{5}{6} - 0 = \frac{5}{6}$
2. Subtract the whole numbers. $5 - 2 = 3$
3. Combine the two. $3 + \frac{5}{6} = 3\frac{5}{6}$ (The difference)

PROOF: $3\frac{5}{6} + 2 = 5\frac{5}{6}$

SUBTRACTING A MIXED NUMBER FROM A WHOLE NUMBER

$$5 - 2\frac{5}{6} =$$

The minuend does not have a fraction.
REMEMBER: You cannot subtract from 0.

1. BORROW 1 from 4 and change it into 6/6 and our problem will look like this:

$$5 - 2\frac{5}{6} = 4\frac{6}{6} - 2\frac{5}{6} =$$

2. Now we can subtract.

$$\frac{6}{6} - \frac{5}{6} = \frac{6-5}{6} = \frac{1}{6}$$

3. Subtract the whole numbers. $4 - 2 = 2$
4. Combine the two. $2 + \frac{1}{6} = 2\frac{1}{6}$ (The difference)

PROOF: $2\frac{1}{6} + 2\frac{5}{6} = 5$

INTERESTING PROBLEM

$$4\frac{1}{6} - 2\frac{5}{6} = \left[3\frac{6}{6} + \frac{1}{6}\right] - 2\frac{5}{6} = 3\frac{7}{6} - 2\frac{5}{6} = 1\frac{2}{6}$$

$$= 1\frac{1}{3} \text{ (The difference)}$$

1. You cannot take 5/6 from 1/6.
2. BORROW 1 and change it to 6/6.
3. Add it to 1/6 to get 7/6.
4. Subtract and reduce to lowest terms.

PROOF: $1\frac{2}{6} + 2\frac{5}{6} = 3\frac{7}{6} = 4\frac{1}{6}$

TIME LIMIT—4 Minutes

SUBTRACTION OF FRACTIONS

$\frac{5}{8} - \frac{1}{8}$	$\frac{4}{7} - \frac{3}{7}$	$\frac{5}{6} - \frac{5}{6}$	$\frac{7}{9} - \frac{4}{9}$	$\frac{9}{10} - \frac{3}{10}$
$1\frac{7}{8} - 1\frac{3}{8}$	$2\frac{3}{7} - 1\frac{2}{7}$	$3\frac{3}{5} - 1\frac{2}{5}$	$4\frac{5}{6} - 1\frac{3}{6}$	$5\frac{4}{9} - 1\frac{2}{9}$
$5\frac{7}{8} - \frac{5}{8}$	$4\frac{5}{7} - \frac{3}{7}$	$7\frac{4}{5} - \frac{2}{5}$	$6\frac{2}{6} - \frac{1}{6}$	$11\frac{5}{9} - \frac{3}{9}$
$5\frac{5}{8} - \frac{7}{8}$	$4\frac{3}{7} - \frac{5}{7}$	$7\frac{2}{5} - \frac{4}{5}$	$6\frac{1}{6} - \frac{2}{6}$	$11\frac{3}{9} - \frac{5}{9}$
$8\frac{1}{7} - 3\frac{5}{7}$	$12\frac{1}{5} - 4\frac{2}{5}$	$9\frac{3}{6} - 6\frac{5}{6}$	$8\frac{4}{8} - 2\frac{7}{8}$	$15\frac{4}{9} - 4\frac{5}{9}$

SCORING

1 error = excellent
2 errors = good

SUBTRACTION OF UNLIKE FRACTIONS

SUBTRACTING A FRACTION FROM A FRACTION

$$\begin{array}{r} 1/2 \\ \underline{-1/3} \end{array}$$

1. The LCD is 6.

2. Change both fractions to their equivalent fractions having a denominator of 6.

$$\frac{1}{2} = \frac{3}{6}$$

$$\frac{1}{3} = \frac{2}{6}$$

Difference 1/6

PROOF: 1/6 + 2/6 = 3/6 = 1/2

SUBTRACTING A FRACTION FROM A MIXED NUMBER

$$\begin{array}{r} 3\ 4/5 \\ \underline{-2/3} \end{array}$$

1. The LCD is 15.

2. Change both fractions to their equivalent fractions having a denominator of 15.

$$\begin{array}{rcr} 3\ 4/5 & = & 3\ 12/15 \\ \underline{-2/3} & = & \underline{10/15} \\ \text{Difference} & & 3\ \ 2/15 \end{array}$$

PROOF:

3 2/15 + 2/3 = 3 2/15 + 10/15 = 3 12/15 = 3 4/5

SUBTRACTING A MIXED NUMBER FROM A MIXED NUMBER

$$\begin{array}{r} 5\ 1/3 \\ \underline{-2\ 5/8} \end{array}$$

1. The LCD is 24.

2. Change both fractions to their equivalent fractions having a denominator of 24.

$$\begin{array}{rcr} 5\ 1/3 & = & 5\ 8/24 \\ \underline{-2\ 5/8} & = & \underline{2\ 15/24} \end{array}$$

3. We cannot subtract 15/24 from 8/24.

4. BORROW 1 from 5 and change it to 24/24.

5. Add the 24/24 to the 8/24 to get 32/24.

$$\begin{array}{rcrcrcr} 5\ 1/3 & = & 5\ \ 8/24 & = & 4\ 24/24 + 8/24 & = & 4\ 32/24 \\ \underline{-2\ 5/8} & = & \underline{2\ 15/24} & = & \underline{2\ 15/24} & = & \underline{2\ 15/24} \end{array}$$

6. Now we can subtract.

$$\begin{array}{rr} & 4\ 32/24 \\ & \underline{-2\ 15/24} \\ \text{Difference} & 2\ 17/24 \end{array}$$

PROOF:

2 17/24 + 2 5/8 = 2 17/24 + 2 15/24 = 4 32/24
4 32/24 = 5 8/24 = 5 1/3

SUBTRACTION OF FRACTIONS

$\frac{13}{14} =$ | $\frac{23}{24} =$ | $\frac{11}{16} =$ | $\frac{13}{15} =$
$-\frac{3}{7} =$ ______ | $-\frac{5}{8} =$ ______ | $-\frac{1}{4} =$ ______ | $-\frac{2}{5} =$ ______

$2\ \frac{5}{8} =$ | $4 =$ | $3\ \frac{7}{8} =$ | $4\ \frac{4}{5} =$
$-\frac{1}{4} =$ ______ | $-\frac{3}{12} =$ ______ | $-\frac{1}{2} =$ ______ | $-\frac{3}{10} =$ ______

$5\ \frac{1}{2} =$ | $7\ \frac{7}{12} =$ | $9\ \frac{7}{8} =$ | $6\ \frac{3}{4} =$
$-2\ \frac{1}{5} =$ ______ | $-3\ \frac{2}{9} =$ ______ | $-4\ \frac{2}{3} =$ ______ | $-3\ \frac{1}{7} =$ ______

$12\ \frac{3}{8} =$ | $15\ \frac{1}{24} =$ | $16\ \frac{3}{7} =$ | $25\ \frac{11}{36} =$
$-5\ \frac{3}{5} =$ ______ | $-7\ \frac{3}{4} =$ ______ | $-3\ \frac{2}{3} =$ ______ | $-6\ \frac{5}{9} =$ ______

SCORING

1 error = excellent
2 errors = good

MULTIPLYING FRACTIONS AND WHOLE NUMBERS

How much is 1/2 of 10 dollars?	5 dollars
How much is 1/2 of 4 dollars?	2 dollars
How much is 1/2 of 1 dollar?	1/2 dollar

How much is 1/3 of 12 dollars?	4 dollars
How much is 1/3 of 6 dollars?	2 dollars
How much is 1/3 of 3 dollars?	1 dollar

How much is 1/4 of 12 dollars?	3 dollars
How much is 1/4 of 4 dollars?	1 dollar
How much is 1/4 of 1 dollar?	1 quarter

If you classify the word problems above as simple, then that's what this lesson will be– simple.

Take the problem 1/5 of 5

We rewrite it as 1/5 x 5

because the word "of" means "to multiply."

It is a well-known fact that in multiplication you may write the factors in any order without affecting the product.

In other words, 1/5 x 5 is the same as 5 x 1/5

1/5 x 5 may be read as 1/5 of 5 = 1
5 x 1/5 may be read as 5 of 1/5 = 1

1/5 + 1/5 + 1/5 + 1/5 + 1/5 = 5/5 = 1

Now that we have gotten through the introductory remarks about this operation, let's get on with the work.

RULE: For multiplying fractions

1. Multiply numerators.
2. Multiply denominators.
3. Reduce all products.

EXAMPLES: When multiplying a whole number by a fraction, we write the whole number over a denominator of 1.
REMEMBER: Any number divided by 1 equals the number.

1. $\frac{1}{5} \times 5 = \frac{1}{5} \times \frac{5}{1} = \frac{5}{5} = 1$

2. $\frac{2}{3} \times 12 = \frac{2}{3} \times \frac{12}{1} = \frac{24}{3} = 8$

3. $\frac{3}{4} \times 9 = \frac{3}{4} \times \frac{9}{1} = \frac{27}{4} = 6\ 3/4$

4. $6 \times \frac{3}{7} = \frac{6}{1} \times \frac{3}{7} = \frac{18}{7} = 2\ 4/7$

OBSERVATION:

Most people firmly believe that the word "multiply" means to increase, and will not accept the fact that this is not necessarily true with fractions.

TIME LIMIT—5 Minutes

MULTIPLYING FRACTIONS

$4 \times \frac{5}{7} =$	$6 \times \frac{2}{9} =$	$8 \times \frac{4}{5} =$	$10 \times \frac{5}{12} =$
$6 \times \frac{7}{8} =$	$9 \times \frac{5}{11} =$	$12 \times \frac{5}{6} =$	$15 \times \frac{7}{10} =$
$8 \times \frac{5}{9} =$	$12 \times \frac{3}{5} =$	$16 \times \frac{5}{7} =$	$20 \times \frac{4}{9} =$
$10 \times \frac{9}{11} =$	$15 \times \frac{8}{13} =$	$20 \times \frac{7}{12} =$	$25 \times \frac{7}{15} =$
$12 \times \frac{15}{16} =$	$18 \times \frac{3}{5} =$	$24 \times \frac{7}{11} =$	$30 \times \frac{7}{20} =$

SCORING

1 error = excellent
2 errors = good

MULTIPLYING A FRACTION BY A FRACTION

How much is 1/2 of a 1/2 dollar? 1 quarter

1/2 of 1/2 = 1/4

In arithmetic the word "of" is replaced by the multiplication sign, x.

$$\frac{1}{2} \times \frac{1}{2} = \frac{1}{4}$$

We are able to get a product of 1/4 by:

1. Multiplying the numerators 1 x 1 = 1
2. Multiplying the denominators 2 x 2 = 4

The most confusing part of multiplying a fraction by a fraction is that all answers turn out to be smaller than the factors.

EXAMPLE:

How much is 1/2 of 2/3 of a pie?

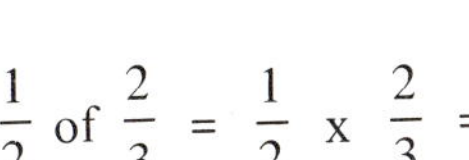

$$\frac{1}{2} \text{ of } \frac{2}{3} = \frac{1}{2} \times \frac{2}{3} =$$

1. Multiply numerators. 1 x 2 = 2
2. Multiply denominators. 2 x 3 = 6

$$\frac{1}{2} \times \frac{2}{3} = \frac{2}{6}$$

3. Reduce the answer. $\frac{2}{6} = \frac{1}{3}$

EXAMPLE:

How much is 2/3 of 2/3?

1/3	1/3	1/3
		7
		8
5	6	9

$$\frac{2}{3} \text{ of } \frac{2}{3} = \frac{2}{3} \times \frac{2}{3}$$

1. Multiply numerators. 2 x 2 = 4
2. Multiply denominators. 3 x 3 = 9

$$\frac{2}{3} \times \frac{2}{3} = \frac{4}{9}$$

EXAMPLE:

How much is 2/3 of 3/4?

1/4	1/4	1/4	1/4
1	2	3	10
4	5	6	11
7	8	9	12

$$\frac{2}{3} \text{ of } \frac{3}{4} = \frac{2}{3} \times \frac{3}{4} =$$

1. Multiply numerators. 2 x 3 = 6
2. Multiply denominators. 3 x 4 = 12

$$\frac{2}{3} \times \frac{3}{4} = \frac{6}{12}$$

3. Reduce the answer. $\frac{6}{12} = \frac{1}{2}$

TIME LIMIT—8 Minutes

MULTIPLYING FRACTIONS

Always reduce the answer to lowest terms.

$\frac{4}{7} \times \frac{5}{6} =$

$\frac{8}{9} \times \frac{13}{20} =$

$\frac{18}{25} \times \frac{5}{8} =$

$\frac{32}{45} \times \frac{9}{16} =$

$\frac{2}{3} \times \frac{9}{11} =$

$\frac{5}{8} \times \frac{3}{5} =$

$\frac{12}{17} \times \frac{11}{18} =$

$\frac{19}{21} \times \frac{7}{8} =$

$\frac{5}{12} \times \frac{3}{7} =$

$\frac{7}{10} \times \frac{5}{9} =$

$\frac{27}{35} \times \frac{7}{9} =$

$\frac{12}{17} \times \frac{9}{20} =$

$\frac{9}{12} \times \frac{7}{18} =$

$\frac{28}{35} \times \frac{5}{7} =$

$\frac{7}{12} \times \frac{9}{15} =$

$\frac{25}{32} \times \frac{8}{13} =$

$\frac{6}{10} \times \frac{12}{15} =$

$\frac{15}{21} \times \frac{7}{9} =$

$\frac{21}{32} \times \frac{2}{3} =$

$\frac{11}{20} \times \frac{5}{7} =$

SCORING

1 error = excellent
2 errors = good

MULTIPLYING A MIXED NUMBER BY A FRACTION

What is 1/2 of 2 1/2 dollars? 1 1/4 dollars

$$\frac{1}{2} \times 2\ 1/2 = 1\ 1/4$$

1. Always change the mixed number to an improper fraction.

$$2\frac{1}{2} = \left[\frac{2 \times 2}{2} + \frac{1}{2}\right] = \frac{5}{2}$$

2. Multiply.

$$\frac{1}{2} \times 2\ 1/2 = \frac{1}{2} \times \frac{5}{2} = \frac{5}{4}$$

3. Change the product to a mixed number.

$$\frac{5}{4} = 1\ 1/4$$

EXAMPLE: What is 2/3 of 4 4/5?

1. Always change the mixed number to an improper fraction.

$$4\frac{4}{5} = \left[\frac{5 \times 4}{5} + \frac{4}{5}\right] = \frac{24}{5}$$

2. Multiply.

$$\frac{2}{3} \times 4\frac{4}{5} = \frac{2}{3} \times \frac{24}{5} = \frac{48}{15}$$

3. Change the product to a mixed number.

$$\frac{48}{15} = 3\ 3/15$$

4. Reduce the fraction to its lowest terms.

$$3\ 3/15 = 3\ 1/5$$

MULTIPLYING A MIXED NUMBER BY A WHOLE NUMBER

What is 3 times 2 1/2 dollars? 7 1/2 dollars

$$3 \times 2\ 1/2 = 7\ 1/2$$

1. Always change the mixed number to an improper fraction.

$$2\frac{1}{2} = \left[\frac{2 \times 2}{2} + \frac{1}{2}\right] = \frac{5}{2}$$

2. Multiply.

$$3 \times \frac{5}{2} = \frac{3}{1} \times \frac{5}{2} = \frac{15}{2}$$

3. Change the product to a mixed number.

$$\frac{15}{2} = 7\ 1/2$$

EXAMPLE: 5 x 5 3/8

$$5 \times \frac{43}{8} = \frac{5}{1} \times \frac{43}{8} = \frac{215}{8} = 26\ 7/8$$

MULTIPLYING A MIXED NUMBER BY A MIXED NUMBER

What is 1/2 of 2 1/2 dollars? 1 1/4 dollars
What is 3 times 2 1/2 dollars? 7 1/2 dollars
What is 3 1/2 times 2 1/2 dollars? 8 3/4 dollars

1. Change both mixed numbers to improper fractions.

$$3\frac{1}{2} = \left[\frac{3 \times 2}{2} + \frac{1}{2}\right] = 7/2$$

$$2\frac{1}{2} = \left[\frac{2 \times 2}{2} + \frac{1}{2}\right] = 5/2$$

2. Multiply.

$$\frac{7}{2} \times \frac{5}{2} = \frac{35}{4} = 8\ 3/4$$

EXAMPLE: 12 2/3 x 10 2/7

$$\frac{38}{3} \times \frac{72}{7} = \frac{2{,}736}{21} = 130\ 6/21 = 130\ 2/7$$

TIME LIMIT—15 Minutes

MULTIPLYING FRACTIONS

2 3/8 x 2/3 =

7 3/4 x 3/5 =

12 1/8 x 5/6 =

36 2/3 x 4/5 =

25 3/7 x 1/4 =

16 3/9 x 2/7 =

21 6/10 x 3/8 =

6 5/9 x 12 =

5 2/7 x 15 =

15 3/10 x 8 =

9 3/16 x 10 =

7 5/6 x 15 =

22 4/9 x 6 =

12 7/10 x 8 =

2 3/8 x 6 2/9 =

7 3/4 x 5 1/7 =

16 2/3 x 3 3/16 =

15 5/7 x 8 1/6 =

5 3/5 x 12 5/10 =

8 1/10 x 14 2/7 =

6 2/9 x 23 1/7 =

SCORING

1 error = excellent
2 errors = good

DIVIDING FRACTIONS

How many quarters in 2 dollars? 8
Therefore 2 ÷ 1/4 must equal 8

How many quarters in 1 1/2 dollars? 6
Therefore 1 1/2 ÷ 1/4 must equal 6

How many quarters in 1/2 dollar? 2
Therefore 1/2 ÷ 1/4 must equal 2

If you classify the problems above as simple—
then that's what this lesson is going to be—SIMPLE.

First, let's review a little bit.

Multiply: $3 \times 4 = 12$

Divide: $3\overline{)\,12}$ = 4

Obviously, multiplication and division are closely related. One is the reverse of the other.
Another word for "reverse" is "INVERSE."

We propose to solve division problems involving fractions by turning them into multiplication problems involving fractions.

EXAMPLE: 2 ÷ 1/4 =

which may be rewritten as $\dfrac{2}{\frac{1}{4}} =$

2 is the numerator
1/4 is the denominator

What will it take to change the denominator of 1/4 to 1?
Multiplying it by its INVERSE.
What is the inverse of 1/4? 4/1
If we multiply 1/4 by 4/1, we will get a product of 1.

But you can't do that to the denominator alone and get away with it. Why not?
Because that will change the value of the number.

What can we do to prevent that?
Multiply the numerator by the same number.

REMEMBER: $\dfrac{\frac{4}{1}}{\frac{4}{1}}$ is another symbol for 1,

and any number multiplied or divided by 1 does not change the value of the number.

SOLUTION

$$\frac{(2 \times 4/1)}{(1/4 \times 4/1)} = \frac{2 \times 4/1}{1} = 2 \times \frac{4}{1} = 8$$

Let us simplify the process above into a rule.

DIVIDING A WHOLE NUMBER BY A FRACTION

RULE:
1. Invert the denominator.
2. Multiply.

EXAMPLE: 5 ÷ 1/2 =

$$5 \times \frac{2}{1} = 10$$

15 ÷ 3/4

$$15 \times \frac{4}{3} = \frac{60}{3} = 20$$

DIVIDING A FRACTION BY A FRACTION

How many 1/4s are there in 1/2? 2

1/2 ÷ 1/4 =

RULE: Invert the denominator and multiply.

$$\frac{1}{2} \times \frac{4}{1} = \frac{4}{2} = 2$$

EXAMPLE: $7/8 \div 3/16 = \dfrac{7}{8} \times \dfrac{16}{3} = \dfrac{112}{24} = 4\ 16/24$

$= 4\ 2/3$

TIME LIMIT–6 Minutes

DIVIDING FRACTIONS

Reduce answers to lowest terms.

5 ÷ 2/5 =	4 ÷ 2/3 =	6 ÷ 3/7 =	8 ÷ 2/9 =
12 ÷ 5/6 =	15 ÷ 7/8 =	21 ÷ 4/9 =	33 ÷ 5/7 =
5/8 ÷ 1/8 =	4/6 ÷ 2/6 =	2/3 ÷ 2/9 =	12/16 ÷ 1/4 =
7/5 ÷ 2/3 =	11/12 ÷ 3/4 =	15/16 ÷ 1/2 =	18/25 ÷ 3/5 =
2/3 ÷ 7/8 =	3/4 ÷ 11/12 =	1/2 ÷ 15/16 =	3/5 ÷ 18/25 =

SCORING

1 error = excellent
2 errors = good

DIVIDING A MIXED NUMBER BY A FRACTION

EXAMPLE: How many quarters in 1 1/2 dollars? 6

$$1\ 1/2 \div 1/4 = \frac{1\ 1/2}{1/4}$$

1. Convert the mixed number to an improper fraction.

$$1\ 1/2 = \left[\frac{2 \times 1}{2} + \frac{1}{2}\right] = \frac{3}{2}$$

The problem rewritten is:

$$1\ 1/2 \div 1/4 = 3/2 \div 1/4$$

2. Invert the denominator and multiply.

$$3/2 \div 1/4 = \frac{3/2}{1/4} = \frac{3}{2} \times \frac{4}{1} = \frac{12}{2} = 6$$

EXAMPLE: How many 1/3s in 2 2/3? 8

$$2\ 2/3 \div 1/3 = \frac{2\ 2/3}{1/3}$$

$$2\ 2/3 = 8/3$$

$$2\ 2/3 \div 1/3 = \frac{8/3}{1/3}$$

$$\frac{8/3}{1/3} = \frac{8}{3} \times \frac{3}{1} = \frac{24}{3} = 8$$

EXAMPLE: How many 5/8s in 4 1/2?

$$4\ 1/2 \div 5/8 = \frac{4\ 1/2}{5/8}$$

$$4\ 1/2 = 9/2$$

$$4\ 1/2 \div 5/8 = \frac{9/2}{5/8}$$

$$\frac{9/2}{5/8} = \frac{9}{2} \times \frac{8}{5} = \frac{72}{10} = 7\ 2/10 = 7\ 1/5$$

DIVIDING A FRACTION BY A WHOLE NUMBER

How many 2 dollar bills can we get out of a quarter?
Sounds crazy doesn't it? We can't even get one.
Yet it's a legitimate question.
If we reword the question and ask,
"A quarter is what part of a $2 bill?"
that would sound okay, wouldn't it? Well,
they both mean the same thing.

EXAMPLE: How many 2s can we get out of 1/4?

$$1/4 \div 2 =$$

When dividing a fraction by a whole number, we write the whole number as a fraction with a denominator of 1.

$$1/4 \div 2 = 1/4 \div 2/1 = \frac{1/4}{2/1}$$

Invert the denominator and multiply.

$$\frac{1/4}{2/1} = \frac{1 \times 1}{4 \times 2} = 1/8$$

EXAMPLE: How many 3s can we get out of 1/3?

$$1/3 \div 3 = 1/3 \div 3/1 = \frac{1/3}{3/1}$$

$$\frac{1/3}{3/1} = \frac{1}{3} \times \frac{1}{3} = 1/9$$

EXAMPLE: How many 9s can we get out of 3/16?

$$3/16 \div 9 = 3/16 \div 9/1 = \frac{3/16}{9/1}$$

$$\frac{3/16}{9/1} = \frac{3}{16} \times \frac{1}{9} = \frac{3}{144} = 1/48$$

TIME LIMIT–6 Minutes

DIVIDING FRACTIONS

Reduce answers to lowest terms.

$3\ 1/2 \div 1/2 =$	$4\ 3/4 \div 1/4 =$	$5\ 2/3 \div 1/3 =$	$6\ 4/5 \div 1/5 =$
$3\ 1/3 \div 2/3 =$	$4\ 3/4 \div 2/3 =$	$5\ 2/3 \div 2/3 =$	$6\ 4/5 \div 2/3 =$
$12\ 3/5 \div 7/10 =$	$15\ 2/3 \div 5/6 =$	$18\ 3/4 \div 3/8 =$	$16\ 5/9 \div 2/3 =$
$2/3 \div 3 =$	$4/5 \div 5 =$	$5/6 \div 4 =$	$7/8 \div 2 =$
$3/8 \div 4 =$	$2/3 \div 8 =$	$2/5 \div 12 =$	$3/4 \div 9 =$

SCORING

1 error = excellent
2 errors = good

DIVIDING A MIXED NUMBER BY A WHOLE NUMBER

How many 2s can we get out of 4?	2
How many 2s can we get out of 3?	1 1/2
How many 2s can we get out of 1/2?	1/4
How many 2s can we get out of 3 1/2?	1 3/4

EXAMPLE: 3 1/2 ÷ 2 =

1. Convert the mixed number to an improper fraction.

3 1/2 ÷ 2 = 7/2 ÷ 2 =

2. Give the whole number a denominator of 1.

$$7/2 \div 2 = 7/2 \div 2/1 = \frac{7/2}{2/1}$$

3. Invert the denominator and multiply.

$$\frac{7/2}{2/1} = \frac{7}{2} \times \frac{1}{2} = \frac{7}{4} = 1\ 3/4$$

EXAMPLE: How many 3s in 11 2/3?

$$11\ 2/3 \div 3 = \frac{11\ 2/3}{3/1} = \frac{35/3}{3/1} =$$

$$\frac{35/3}{3/1} = \frac{35}{3} \times \frac{1}{3} = \frac{35}{9} = 3\ 8/9$$

EXAMPLE: How many 5s in 16 1/4?

$$16\ 1/4 \div 5 = \frac{16\ 1/4}{5/1} = \frac{65/4}{5/1} =$$

$$\frac{65/4}{5/1} = \frac{65}{4} \times \frac{1}{5} = \frac{65}{20} = 3\ 15/20 = 3\ 1/4$$

DIVIDING A MIXED NUMBER BY A MIXED NUMBER

How many 2 1/2s in 2 1/2?	1
How many 2 1/2s in 5?	2
How many 2 1/2s in 7 1/2?	3

EXAMPLE: 7 1/2 ÷ 2 1/2 =

1. Convert both mixed numbers to improper fractions.

7 1/2 ÷ 2 1/2 = 15/2 ÷ 5/2

2. Invert and multiply.

$$15/2 \div 5/2 = \frac{15/2}{5/2} = \frac{15}{2} \times \frac{2}{5} = \frac{30}{10} = 3$$

EXAMPLE: How many 2 2/3s in 7 1/9?

$$7\ 1/9 \div 2\ 2/3 = \frac{7\ 1/9}{2\ 2/3}$$

$$= \frac{64/9}{8/3} = \frac{64}{9} \times \frac{3}{8} = \frac{192}{72} = 2\ 48/72 = 2\ 2/3$$

EXAMPLE: How many 6 3/8s in 15 3/5?

$$15\ 3/5 \div 6\ 3/4 = \frac{15\ 3/5}{6\ 3/4}$$

$$\frac{15\ 3/5}{6\ 3/4} = \frac{78/5}{27/4} = \frac{78}{5} \times \frac{4}{27} = \frac{312}{135} = 2\ 42/135 = 2\ 14/45$$

TIME LIMIT—15 Minutes

DIVIDING FRACTIONS

Reduce answers to lowest terms.

10 1/8 ÷ 4 =	12 1/7 ÷ 2 =	16 4/5 ÷ 3 =	13 1/8 ÷ 5 =
27 3/4 ÷ 6 =	15 3/8 ÷ 7 =	22 5/6 ÷ 6 =	18 2/3 ÷ 8 =
12 1/7 ÷ 7 1/2 =	10 1/8 ÷ 4 1/2 =	13 1/8 ÷ 5 1/4 =	27 2/4 ÷ 4 1/8 =
7 1/2 ÷ 12 1/7 =	4 1/2 ÷ 10 1/8 =	5 1/4 ÷ 13 1/8 =	4 1/8 ÷ 27 2/4 =
35 1/5 ÷ 5 1/7 =	44 2/3 ÷ 8 3/8 =	51 4/6 ÷ 9 3/8 =	27 3/4 ÷ 6 1/8 =

SCORING

1 error = excellent
2 errors = good

How many 1 1/4s can we get out of a quarter?

Reworded: 1/4 is what part of 1 1/4? 1/5

EXAMPLE: 1/4 ÷ 1 1/4

1. Convert the mixed number to an improper fraction.

$$1/4 \div 1\ 1/4 = 1/4 \div 5/4 = \frac{1/4}{5/4}$$

2. Invert and multiply.

$$\frac{1/4}{5/4} = \frac{1}{4} \times \frac{4}{5} = \frac{4}{20} = 1/5$$

EXAMPLE: How many 3 1/2s can we get out of 1/2?

Reworded: 1/2 is what part of 3 1/2?

1/2 ÷ 3 1/2 =

$$1/2 \div 3\ 1/2 = 1/2 \div 7/2 = \frac{1/2}{7/2}$$

$$\frac{1/2}{7/2} = \frac{1}{2} \times \frac{2}{7} = \frac{2}{14} = 1/7$$

EXAMPLE: How many 6 2/5s can we get out of 4/12?

Reworded: 4/12 is what part of 6 2/15?

4/12 ÷ 6 2/15?

$$4/12 \div 6\ 2/15 = 4/12 \div 92/15 = \frac{4/12}{92/15}$$

$$\frac{4/12}{92/15} = \frac{4}{12} \times \frac{15}{92} = 60/1104 = 5/92$$

DIVIDING A FRACTION BY A MIXED NUMBER

How many 2 1/2s can we get out of 10? 4

EXAMPLE: 10 ÷ 2 1/2

1. Convert the mixed number to an improper fraction.

$$10 \div 2\ 1/2 = 10 \div 5/2$$

2. Write the whole number as a fraction with a denominator of 1.

$$10/1 \div 5/2 = \frac{10/1}{5/2}$$

3. Invert the denominator and multiply.

$$\frac{10/1}{5/2} = \frac{10}{1} \times \frac{2}{5} = \frac{20}{5} = 4$$

EXAMPLE: How many 3 1/3s can we get out of 10?

10 ÷ 3 1/3

$$10 \div 3\ 1/3 = 10/1 \div 10/3 = \frac{10/1}{10/3} =$$

$$\frac{10/1}{10/3} = \frac{10}{1} \times \frac{3}{10} = \frac{30}{10} = 3$$

EXAMPLE: How many 1 7/8s can we get out of 12?

12 ÷ 1 7/8 =

$$12 \div 1\ 7/8 = 12/1 \div 15/8 = \frac{12/1}{15/8} =$$

$$\frac{12/1}{15/8} = \frac{12}{1} \times \frac{8}{15} = \frac{96}{15} = 6\ 6/15 = 6\ 2/5$$

TIME LIMIT–10 Minutes

DIVIDING FRACTIONS

Reduce answers to lowest terms.

1/2 ÷ 10 1/8 =	4/5 ÷ 2 9/10 =	4/9 ÷ 1 4/8 =	1/7 ÷ 7 1/2 =
7/12 ÷ 12 1/2 =	3/8 ÷ 8 1/8 =	9/10 ÷ 2 4/7 =	3/5 ÷ 4 7/10 =
1/4 ÷ 13 1/8 =	5/14 ÷ 12 1/7 =	4/5 ÷ 8 1/3 =	3/4 ÷ 4 1/8 =
10 ÷ 1 5/8 =	12 ÷ 2 3/4 =	15 ÷ 3 3/5 =	16 ÷ 5 3/8 =
13 ÷ 5 1/4 =	16 ÷ 2 4/7 =	12 ÷ 7 1/2 =	10 ÷ 4 1/2 =
9 ÷ 2 5/6 =	18 ÷ 3 9/10 =	17 ÷ 2 7/9 =	21 ÷ 5 3/5 =

SCORING

1 error = excellent
2 errors = good

CANCELLATION–THE USE OF FACTORS TO SHORTEN THE WORK

Numbers involved in a multiplication operation are called FACTORS.
Division is the inverse of multiplication.
Numbers involved in a division operation may also be called FACTORS.

Factors separated by a fraction "bar" may be reduced to lower terms, providing they have a "common divisor."

REMEMBER: Dividing the numerator and denominator by the same number does not change the value of the fraction.

Some textbooks refer to this process as CANCELLATION.

EXAMPLE: $\frac{4 \times 2}{2} = \frac{8}{2} = 4$

By cancellation: Dividing both numerator and denominator by 2.

$$\frac{4 \times \not{2}^{1}}{\not{2}_{1}} = 4$$

Both answers do agree.

CAUTION: You cannot cancel if the numbers in the numerator or denominator are involved in addition or subtraction.

EXAMPLE OF PROPER PROCEDURE:

$$\frac{4+2}{2} = \frac{4}{2} + \frac{2}{2} = \frac{6}{2} = 3$$

EXAMPLE OF IMPROPER PROCEDURE:

$$\frac{4 + \not{2}^{1}}{\not{2}_{1}} = \frac{5}{1} = 5$$

BE ON THE ALERT AND SPOT THE FACTORS WHICH CAN BE REDUCED.

EXAMPLE 1: $\frac{6 \times 2}{2 \times 3} = \frac{6 \times \not{2}^{1}}{\not{2}_{1} \times 3} =$

1. Divide both numerator and denominator by 2.
2. Divide both numerator and denominator by 3.

$$\frac{\not{6}^{2} \times \not{2}^{1}}{\not{2}_{1} \times \not{3}_{1}} = 2$$

EXAMPLE 2: $\frac{15 \times 12 \times 7}{3 \times 5} = \frac{\not{15}^{3} \times 12 \times 7}{3 \times \not{5}_{1}} =$

1. Divide both numerator and denominator by 5.
2. Divide both numerator and denominator by 3.

$$\frac{\not{15}^{\not{3}\,1} \times 12 \times 7}{\not{3}_{1} \times \not{5}_{1}} = 84$$

BY THE WAY, the division above may be performed in any order

EXAMPLE 3: $\frac{35 \times 16 \times 21 \times 12}{15 \times 3 \times 7}$

1. Divide $\frac{21}{7}$ by $\frac{7}{7}$. $\quad \frac{35 \times 16 \times \not{21}^{3} \times 12}{15 \times 3 \times \not{7}_{1}}$
2. Divide $\frac{3}{3}$ by $\frac{3}{3}$. $\quad \frac{35 \times 16 \times \not{21}^{\not{3}\,1} \times 12}{15 \times \not{3}_{1} \times \not{7}_{1}}$
3. Divide $\frac{35}{15}$ by $\frac{5}{5}$. $\quad \frac{\not{35}^{7} \times 16 \times \not{21}^{\not{3}\,1} \times 12}{\not{15}_{3} \times \not{3}_{1} \times \not{7}_{1}}$
4. Divide $\frac{12}{3}$ by $\frac{3}{3}$. $\quad \frac{\not{35}^{7} \times 16 \times \not{21}^{\not{3}\,1} \times \not{12}^{4}}{\not{15}_{\not{3}\,1} \times \not{3}_{1} \times \not{7}_{1}}$
5. The answer is the product of the remaining factors. $\quad 7 \times 16 \times 4 = 448$

Well, what have you got to say?
It's much easier this way, isn't it?

CANCELLATION OF FRACTIONS

Reduce answers to lowest terms.

$\frac{3 \times 2}{6 \times 9} =$ $\frac{5 \times 2}{8 \times 10} =$ $\frac{4 \times 21}{7 \times 30} =$ $\frac{7 \times 12}{8 \times 21} =$

$\frac{3 \times 8 \times 5}{4} =$ $\frac{4 \times 7 \times 6}{3} =$ $\frac{5 \times 8 \times 9}{6} =$ $\frac{7 \times 12 \times 15}{9} =$

$\frac{14 \times 12 \times 16}{4 \times 7} =$ $\frac{15 \times 18 \times 21}{3 \times 5} =$ $\frac{10 \times 24 \times 8}{5 \times 3} =$ $\frac{5 \times 27 \times 30}{10 \times 15} =$

$\frac{64 \times 7 \times 21}{3 \times 8 \times 4} =$ $\frac{36 \times 48 \times 51}{17 \times 12 \times 9} =$ $\frac{25 \times 16 \times 81}{9 \times 64 \times 15} =$ $\frac{24 \times 49 \times 63}{9 \times 14 \times 8} =$

$\frac{121 \times 72 \times 15 \times 9}{9 \times 25 \times 11} =$ $\frac{45 \times 38 \times 80 \times 144}{24 \times 57 \times 120} =$

$\frac{180 \times 62 \times 52 \times 30}{8 \times 15 \times 52 \times 39} =$ $\frac{65 \times 78 \times 135 \times 49}{14 \times 13 \times 39 \times 35} =$

SCORING

1 error = excellent
2 errors = good

FRACTIONS–MULTIPLE OPERATIONS

The problems that follow contain combinations of addition, subtraction, multiplication and division of fractions. Before solving such problems, you must remember the ORDER the solving process must take.

ORDER OF OPERATIONS:

1. Simplify the numerator and denominator
   a. Do all multiplication and division first, in the order in which they appear FROM LEFT TO RIGHT.
   b. Do all addition and subtraction in the order in which they appear FROM LEFT TO RIGHT.
2. Solve all multiplication and division operations.
3. Last, do all additions and subtractions.

EXAMPLE 1:

$$\frac{5/6 + 1/3}{2/3} =$$

1. Simplify the numerator.

$$5/6 + 1/3 = 5/6 + 2/6 = 7/6$$

2. The problem has been reduced to: $\frac{7/6}{2/3}$

3. Invert the denominator and multiply. Reduce by cancellation.

$$\frac{7}{\cancel{6}_2} \times \frac{\cancel{3}^1}{2} = \frac{7}{4} = 1\ 3/4$$

EXAMPLE 2:

$$\frac{2/3}{5/6 + 1/3} =$$

1. Simplify the denominator. (See Example 1.) $= 7/6$

2. Invert the denominator and multiply. Reduce by cancellation.

$$\frac{2}{\cancel{3}_1} \times \frac{\cancel{6}^2}{7} = 4/7$$

EXAMPLE 3:

$$\frac{7\ 1/2 - 3\ 3/7}{3\ 3/4} =$$

1. Always change mixed numbers to improper fractions.

$$\frac{15/2 - 24/7}{15/4} =$$

2. Simplify the numerator.

$$15/2 - 24/7 = 105/14 - 48/14 = 57/14$$

3. The problem simplified is: $\frac{57/14}{15/4}$

4. Invert the denominator and multiply. Reduce by cancellation.

$$\frac{\cancel{57}^{19}}{\cancel{14}_7} \times \frac{\cancel{4}^2}{\cancel{15}_5} = \frac{38}{35} = 1\ 3/35$$

EXAMPLE 4:

$$\frac{3/4 \times 5/6}{3/8} =$$

1. We could simplify the numerator first—or invert the denominator and multiply all three fractions at the same time.

2. Always reduce by cancellation.

$$\frac{\cancel{3}^1}{\cancel{4}_1} \times \frac{5}{\cancel{6}_3} \times \frac{\cancel{8}^{\cancel{2}^1}}{3} = \frac{5}{3} = 1\ 2/3$$

EXAMPLE 5:

$$\frac{3/8}{3/4 \times 5/6} =$$

1. We could simplify the denominator first—or invert the entire denominator (only because it is a multiplication operation) and multiply.

2. Always reduce by cancellation.

$$\frac{\cancel{3}^1}{\cancel{8}_{\cancel{2}_1}} \times \frac{\cancel{4}^1}{\cancel{3}_1} \times \frac{\cancel{6}^3}{5} = \frac{3}{5}$$

TIME LIMIT–20 Minutes

MULTIPLE OPERATIONS–FRACTIONS

Reduce answers to lowest terms.

$$\frac{2/3 + 3/4}{3/5} = \qquad \frac{3/7 + 5/8}{5/6} = \qquad \frac{4/9 + 4/5}{3/7} = \qquad \frac{5/6 + 3/7}{3/4} =$$

$$\frac{2/3}{3/4 + 3/5} = \qquad \frac{3/7}{5/8 + 5/6} = \qquad \frac{4/9}{4/5 + 3/7} = \qquad \frac{5/6}{3/7 + 3/4} =$$

$$\frac{3\ 3/4 - 2\ 2/3}{4\ 1/6} = \qquad \frac{5\ 7/8 - 1\ 2/3}{2\ 1/7} = \qquad \frac{6\ 3/8 - 2\ 3/4}{3\ 1/3} = \qquad \frac{7\ 5/9 - 3\ 4/7}{5\ 1/6} =$$

$$\frac{7/8 \times 5/9}{2/3} = \qquad \frac{3/16 \times 4/7}{7/9} = \qquad \frac{4/15 \times 7/8}{5/7} = \qquad \frac{5/32 \times 15/16}{5/8} =$$

$$\frac{7/8}{5/9 \times 2/3} = \qquad \frac{3/16}{4/7 \times 7/9} = \qquad \frac{4/15}{7/8 \times 5/7} = \qquad \frac{5/32}{15/16 \times 5/8} =$$

SCORING

1 error = excellent
2 errors = good

EXAMPLE 6:

$$\frac{4/9 \times 3/7}{4/9 + 3/7} =$$

1. Simplify the denominator.

$$4/9 + 3/7 = 28/63 + 27/63 = 55/63$$

2. The problem simplified is:

$$\frac{4/9 \times 3/7}{55/63}$$

3. Invert the denominator and multiply. Reduce by cancellation.

$$\frac{4}{\cancel{9}_{1}} \times \frac{3}{\cancel{7}_{1}} \times \frac{\cancel{63}^{\cancel{7}\,1}}{55} = \frac{12}{55}$$

EXAMPLE 7:

$$\frac{4/9 + 3/7}{4/9 \times 3/7} =$$

1. Simplify the numerator. $= 55/63$

2. The problem simplified is:

$$\frac{55/63}{4/9 \times 3/7} =$$

3. Invert the entire denominator and multiply. Reduce by cancellation.

$$\frac{55}{\cancel{63}_{\cancel{7}\,1}} \times \frac{\cancel{9}^{1}}{4} \times \frac{\cancel{7}^{1}}{3} = \frac{55}{12} = 4\ 7/12$$

EXAMPLE 8:

$$\frac{3/8 + 1\ 3/6 \times 3/8}{5/6}$$

1. Simplify the numerator. REMEMBER: multiplication comes first.

$$1\ 3/6 \times 3/8 =$$

$$\frac{9}{\cancel{6}_{2}} \times \frac{\cancel{3}^{1}}{8} = \frac{9}{16}$$

Now perform the addition.

$$\frac{9}{16} + \frac{3}{8} = \frac{9}{16} + \frac{6}{16} = \frac{15}{16}$$

2. Invert the denominator and multiply.

$$\frac{\cancel{15}^{3}}{\cancel{16}_{8}} \times \frac{\cancel{6}^{3}}{\cancel{5}_{1}} = \frac{9}{8} = 1\ 1/8$$

EXAMPLE 9:

$$\frac{14\ 2/3 \div 3\ 3/5 - 5/9}{5/9}$$

1. Simplify the numerator. REMEMBER: division comes first.

$$14\ 2/3 \div 3\ 3/5 =$$

2. Always change mixed numbers to improper fractions.

$$\frac{44}{3} \div \frac{18}{5} = \frac{\cancel{44}^{22}}{3} \times \frac{5}{\cancel{18}_{9}} = \frac{110}{27}$$

3. Now subtract.

$$\frac{110}{27} - \frac{5}{9} = \frac{110}{27} - \frac{15}{27} = \frac{95}{27}$$

4. Invert the denominator and multiply.

$$\frac{\cancel{95}^{19}}{\cancel{27}_{3}} \times \frac{\cancel{9}^{1}}{\cancel{5}_{1}} = \frac{19}{3} = 6\ 1/3$$

MULTIPLE OPERATIONS–FRACTIONS

Reduce answers to lowest terms.

$\dfrac{7/8 \times 3/4}{7/8 + 3/4} =$ $\dfrac{7/9 \times 2/3}{7/9 + 2/3} =$ $\dfrac{3/5 \times 5/12}{3/5 + 5/12} =$ $\dfrac{5/6 \times 3/7}{5/6 + 3/7} =$

$\dfrac{3/8 + 7/9}{3/8 \times 7/9} =$ $\dfrac{4/7 + 3/5}{4/7 \times 3/5} =$ $\dfrac{5/16 + 7/8}{5/16 \times 7/8} =$ $\dfrac{7/10 + 5/6}{7/10 \times 5/6} =$

$\dfrac{7/8 + 2\ 3/4 \times 5/6}{3/5} =$ $\dfrac{5/16 + 3\ 4/5 \times 1\ 2/3}{1\ 5/8} =$ $\dfrac{4\ 1/5 \times 3\ 1/3 + 5\ 1/6}{7\ 3/8} =$

$\dfrac{15\ 5/8 \div 3\ 3/4 - 5/6}{7/9} =$ $\dfrac{23\ 1/5 \div 4\ 5/6 - 1\ 7/8}{2\ 3/4} =$ $\dfrac{18\ 2/3 - 8\ 2/3 \div 2\ 1/2}{3\ 3/7} =$

SCORING

1 error = excellent
2 errors = good

Add:

3 3/5	2 3/4	6 3/8	12 5/12
1/5	5 2/3	2 4/5	6 1/6
2 4/5	6 7/8	5 7/10	5 3/4

Subtract:

3 4/5	3 3/5	4 5/7	5 3/16
−2 3/5	−2 4/5	−2 3/5	−2 3/4

Multiply:

7/8 x 5/14 = 5/9 x 3/20 = 5 5/7 x 6 1/8 = 60 x 11/12 =

3 5/8 x 2 4/5 = 9 2/3 x 3 = 3/8 x 4/5 x 5/6 =

Divide:

6 2/3 ÷ 10 = 10 ÷ 6 2/3 = 2/3 ÷ 5/12 =

5/8 ÷ 15/16 = 3 3/8 ÷ 1 5/16 = 3 1/7 ÷ 10 1/2 =

Solve the following:

$\frac{4\ 3/4 \times 4\ 7/8}{6\ 1/2} =$ 8 2/3 + 2 5/8 x 3 3/7 =

TIME LIMIT–6 Minutes

REFRESHER TEST 4

Add:

7	85	479	4,539	5,283
8	97	625	6,108	87
4	86	356	7,869	2,594
9	54	887	8,574	468
6	49	549		
8	68	739		
5	73	964		
9				

Subtract:

72	810	9,003	78,056	304,985
39	336	785	19,278	29,087

Multiply:

15	88	818	705	4,081
18	8	9	38	805

Divide:

8) 125	94) 787	85) 4,530	59) 41,713	47) 141,423

RELATED PROBLEMS–FRACTIONS

Many different things can be thought of in terms of fractions.

AREA is the product of LENGTH times WIDTH.

How many square inches are there in a piece of material with the following dimensions: length, 20 5/6 inches; width, 15 3/5 inches. __________

VOLUME is the product of LENGTH times WIDTH times HEIGHT.

How many cubic inches are there in a piece of material with the following dimensions: length, 12 3/8 inches; width, 11 1/3 inches; height, 8 4/11 inches. __________

WEIGHT If one loaf of bread weighs 1 1/8 pounds, what will be the total weight of 26 loaves? __________

TIME If it took Jones 1/6th of an hour to cut a piece of wood, How long will it take him to cut 50 pieces? __________

MONEY How many quarters in 16 3/4 dollars? __________

GROUPS An order was received for tubing 1 3/4 feet long, tied 6 pieces to a bundle. How many bundles can be made from 126 feet of tubing? __________

If 1 pound of wild rice costs $1.60 (16 ounces equal 1 pound), find the cost of a package weighing 5 ounces. __________

If 1 quart of vinegar costs 64 cents (32 fluid ounces equal 1 quart), find the cost of a bottle of 10 ounces. __________

If 1 pound of steak costs $1.44, find the cost of a steak weighing 1 3/8 pounds. __________

RELATED PROBLEMS–FRACTIONS

Find the total length, in inches, of the piece of wood below.

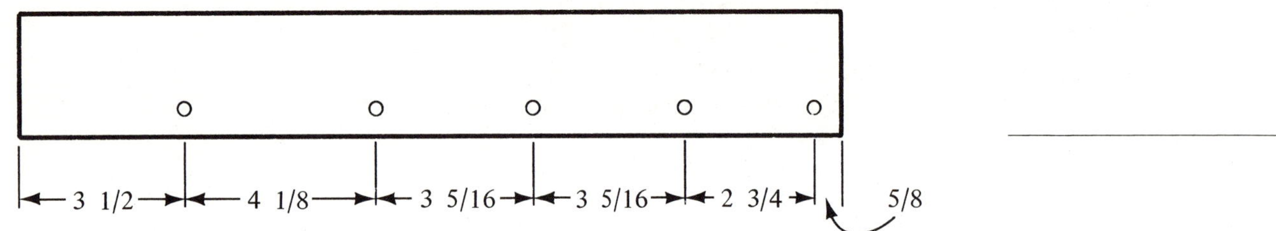

____

From a roll of ribbon 54 3/4 yards long, a piece 8 1/2 yards long is cut
How much ribbon is left on the roll? ________________

The rainfall for five days was as follows:

Monday	3/8 inches
Tuesday	3/16 inches
Wednesday	3/4 inches
Thursday	1 1/8 inches
Friday	1/32 inches

What was the total rainfall for the 5 days? ________________

Mother bought two pieces of roast beef that weighed 6 3/4 pounds together. If one piece weighed 2 7/8 pounds, what was the weight of the other piece? ________________

From a piece of lumber 12 feet long, two pieces were cut. The first was 3 11/16 feet and the other was 7 7/8 feet. How much of the original piece was used? ________________

How much of the original piece is left? ________________

A bottle was 3/4 full. If 2/3 of it was poured out, what part of the full bottle remained? ________________

A man can do a certain job in 5/6 of an hour. A machine can do the same job in 1/2 of this time. How long does it take the machine to do the job? ________________

Chapter 6

DECIMALS

PRETEST–DECIMALS

The addition, subtraction, multiplication, and division of decimals follow the same procedures used with whole numbers, except for two very important things:

1. The positioning of the decimal point

2. The use of zeros as placeholders

Of course, we will assume that by now your errors in computation have been reduced considerably.

Keep the pressure on.

Take the pretest.

Let's find out whether you're on friendly terms with the decimal point.

Is the decimal point that important?

> Don't tell me that you would accept a check for $25.00 when it should read $250.00.

The learning sequence in this chapter will again proceed step by step, introducing only one new element or difficulty at a time.

PRETEST–DECIMALS

Add: 12.05 + .138 + 6.007 + 25 + .0068 =

.0135 + .00079 + .00824 + .000037 + 10 =

Subtract: 12.05 − 3.008 =

.79 − .033 =

Multiply:

.624 × 4	1.257 × 7	27.006 × 6	3601.4 × 9
62.5 × .12	37.05 × .23	6.007 × .42	726.5 × .35

Divide:

12) .6	.12) 144	1.1) 1.21	25) .0625
.0032) 480	2.7) .00108	.18) 1.26	35.3) .25063

DECIMAL FRACTIONS

In our decimal system of notation, we place a value on each column.
Each column has a value that is 1/10 as great as the column to its left.

For example: 1,000,000

1/10	x	1,000,000	=	100,000
1/10	x	100,000	=	10,000
1/10	x	10,000	=	1,000
1/10	x	1,000	=	100
1/10	x	100	=	10
1/10	x	10	=	1

Do numbers less than 1 exist? Of course they do; we just finished a chapter on FRACTIONS.
Our fractions contained a numerator and a denominator.
In this chapter we are going to learn how to name these fractions another way.
They will be called DECIMAL FRACTIONS.
Decimal fractions are special fractions whose denominators are multiples of 10.

For example:

$$\frac{2}{10} \quad \frac{5}{100} \quad \frac{7}{1{,}000} \quad \frac{4}{10{,}000} \quad \frac{3}{100{,}000} \quad \frac{8}{1{,}000{,}000}$$

Decimal fractions may be written without a denominator.

For example:

.2 .05 .007 .0004 .00003 .000008

To learn how to write them, we must extend the place-value system of notation to the right of 1.
To indicate that the number 1 is the last of the whole numbers, we shall use a "dot" or "period" to the right of 1—it shall be called the DECIMAL POINT.

Like this: 1.

To continue the pattern we established above, the new place to the right of the decimal point must have a value equal to 1/10 the value of 1.

For example:

$$\frac{1}{10} \times 1 = \frac{1}{10}$$

This is written as .1

The first place to the right of the decimal point is called the TENTHS PLACE.
Thus .1 is read as one tenth.

Note how the name of the decimal fraction is related to the name of its denominator.

The second place to the right of the decimal point must have a value equal to:

$$\frac{1}{10} \times \frac{1}{10} = \frac{1}{100}$$

This is written as .01

The second place to the right of the decimal point is called the HUNDREDTHS PLACE.
Thus .01 is read as one hundredth.

Note how the name of the decimal fraction is related to the name of its denominator.

The third place to the right of the decimal point must have a value equal to:

$$\frac{1}{10} \times \frac{1}{100} = \frac{1}{1{,}000}$$

This is written as .001

The third place to the right of the decimal point is called the THOUSANDTHS PLACE. Thus .001 is read as one thousandth.

Note how the name of the decimal fraction is related to the name of its denominator.

The fourth place to the right of the decimal point must have a value equal to:

$$\frac{1}{10} \times \frac{1}{1{,}000} = \frac{1}{10{,}000}$$

This is written as .0001

The fourth place to the right of the decimal point is called the TEN THOUSANDTH PLACE. Thus .0001 is read as one ten thousandth.

Note how the name of the decimal fraction is related to the name of its denominator.

The fifth place to the right of the decimal point must have a value equal to:

$$\frac{1}{10} \times \frac{1}{10{,}000} = \frac{1}{100{,}000}$$

This is written as .00001

The fifth place to the right of the decimal point is called the HUNDRED THOUSANDTH PLACE. Thus .00001 is read as one hundred thousandth.

Note how the name of the decimal fraction is related to the name of its denominator.

The sixth place to the right of the decimal point must have a value equal to:

$$\frac{1}{10} \times \frac{1}{100{,}000} = \frac{1}{1{,}000{,}000}$$

This is written as .000001

The sixth place to the right of the decimal point is called the MILLIONTH PLACE. Thus .000001 is read as one millionth.

Note how the name of the decimal fraction is related to the name of its denominator.

A FULL DISPLAY OF THE PLACE-VALUE SYSTEM

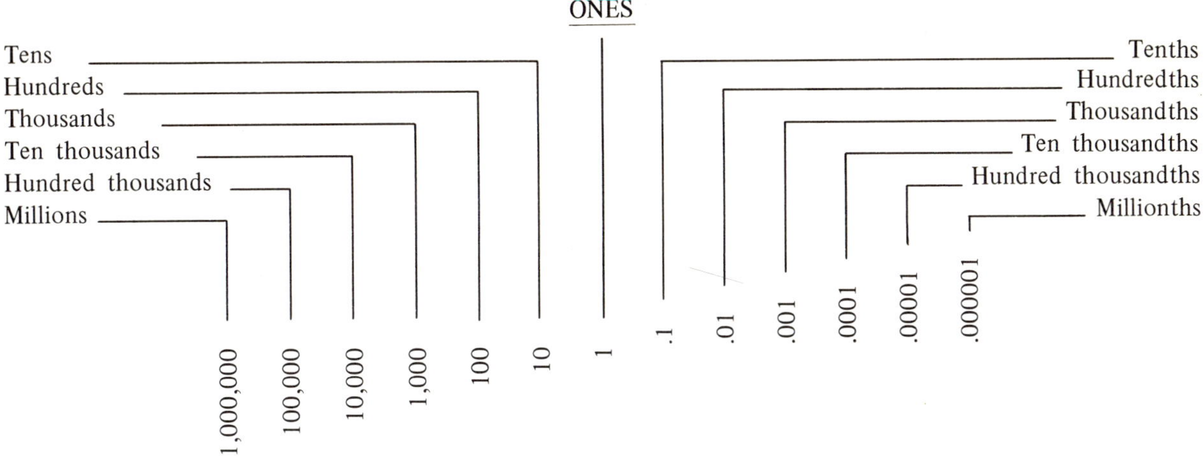

OBSERVATIONS

1. The decimal point is used to separate the whole numbers from the decimal fractions.

2. The decimal point is not the center of the decimal system of notation.

3. The center is the ones place.

4. Note how the TENS are balanced by the TENTHS,
   HUNDREDS by the HUNDREDTHS
   THOUSANDS by the THOUSANDTHS
   TEN THOUSANDS by the TEN THOUSANDTHS
   HUNDRED THOUSANDS by the
   HUNDRED THOUSANDTHS
   MILLIONS by the MILLIONTHS

5. Note how 3/10 and .3 are considered equivalent values. This is true also of 5/100 and .05, 7/1,000 and .007, and so on.

If decimal fractions and fractions are the same, then why bother to learn them?

Because we believe you will find that it is usually easier to ADD, SUBTRACT, MULTIPLY, and DIVIDE decimal fractions than fractions.

READING DECIMALS

The fraction 3/10 is read as	3 TENTHS.
The decimal .3 is read as	3 TENTHS.
The fraction 5/100 is read as	5 HUNDREDTHS.
The decimal .05 is read as	5 HUNDREDTHS.
The fraction 25/100 is read as	25 HUNDREDTHS.
The decimal .25 is read as	25 HUNDREDTHS.
The fraction 2/1,000 is read as	2 THOUSANDTHS.
The decimal .002 is read as	2 THOUSANDTHS.
The fraction 27/1,000 is read as	27 THOUSANDTHS.
The decimal .027 is read as	27 THOUSANDTHS.

OBSERVATION: The number of places to the right of the decimal point equals the number of zeros in the denominator of the fraction that represents the decimal.

SUMMARY:

.5 = 5/10
.05 = 5/100
.005 = 5/1,000

RULE FOR READING DECIMALS:

1. Read the number to the right of the decimal point as you would a whole number.
2. For a denominator, name the place value of the last digit on the right.

EXAMPLE:

.025 Read it as "twenty-five"; then add the place value of the last digit (5).
.025 TWENTY-FIVE THOUSANDTHS

.0183 Read it as "one hundred eighty-three." Then add the place value of the last digit (3).
.0183 ONE HUNDRED EIGHTY-THREE TEN-THOUSANDTHS.

.000724 Read it as "seven hundred twenty-four." Then add the place value of the last digit (4).
.000724 SEVEN HUNDRED TWENTY-FOUR MILLIONTHS.

MIXED DECIMALS 3.5 is called a mixed decimal because it contains a whole number plus a decimal.

It is read as
3 and 5 tenths

The word "and" is used to represent the decimal point.

40.02 is read as 40 and 2 hundredths

WRITING DECIMALS

Nine tenths means 9/10 and is written as .9

Thirty-eight hundredths means 38/100 and is written as .38

Thirteen thousandths means 13/1,000 and is written as .013

Seven hundred thirty-eight hundred-thousandths means 738/100,000 and is written as .00738.

OBSERVATION: The number of places in a decimal is equal to the number of zeros contained in the denominator of the equivalent fraction.

READING FRACTIONS

The fraction 2/10 is read as ______________________

The fraction 3/100 is read as ______________________

The fraction 25/100 is read as ______________________

The fraction 5/1,000 is read as ______________________

The fraction 35/1,000 is read as ______________________

The fraction 455/1,000 is read as ______________________

The fraction 7/10,000 is read as ______________________

The fraction 48/10,000 is read as ______________________

The fraction 155/10,000 is read as ______________________

The fraction 3,525/10,000 is read as ______________________

The fraction 3/100,000 is read as ______________________

The fraction 78/100,000 is read as ______________________

The fraction 229/100,000 is read as ______________________

The fraction 5,575/100,000 is read as ______________________

The fraction 62,123/100,000 is read as ______________________

The fraction 8/1,000,000 is read as ______________________

The fraction 56/1,000,000 is read as ______________________

The fraction 205/1,000,000 is read as ______________________

The fraction 3,602/1,000,000 is read as ______________________

The fraction 47,135/1,000,000 is read as ______________________

The fraction 825,227/1,000,000 is read as ______________________

READING DECIMALS

The decimal .2 is read as ______________________

The decimal .03 is read as ______________________

The decimal .25 is read as ______________________

The decimal .005 is read as ______________________

The decimal .035 is read as ______________________

The decimal .455 is read as ______________________

The decimal .0007 is read as ______________________

The decimal .0048 is read as ______________________

The decimal .0155 is read as ______________________

The decimal .3525 is read as ______________________

The decimal .00003 is read as ______________________

The decimal .00078 is read as ______________________

The decimal .00229 is read as ______________________

The decimal .05575 is read as ______________________

The decimal .62123 is read as ______________________

The decimal .000008 is read as ______________________

The decimal .000056 is read as ______________________

The decimal .000205 is read as ______________________

The decimal .003602 is read as ______________________

The decimal .047135 is read as ______________________

The decimal .825227 is read as ______________________

The decimal 2.04 is read as ______________________

The decimal 13.005 is read as ______________________

WRITING DECIMALS

A cent written as a decimal is ________________

A nickel written as a decimal is ________________

A dime written as a decimal is ________________

Fifteen cents written as a decimal is ________________

Twenty-five cents written as a decimal is ________________

Sixty-five cents written as a decimal is ________________

Ninety cents written as a decimal is ________________

One dollar and five cents written as a decimal is ________________

One dollar and twenty-five cents written as a decimal is ________________

Ten dollars and one cent written as a decimal is ________________

Ten dollars and ten cents written as a decimal is ________________

Fifty dollars and five cents written as a decimal is ________________

Fifty dollars and fifty cents written as a decimal is ________________

One hundred dollars and one cent written as a decimal is ________________

One hundred dollars and ten cents written as a decimal is ________________

ADDING DECIMALS

The basic rule for all addition problems is:

You can add only like values.

quarters to quarters
dimes to dimes
cents to cents
1/2s to 1/2s
1/10s to 1/10s
1/100s to 1/100s and so on.

EXAMPLES: Add

```
  3 tenths
  4 tenths
  --------
  7 tenths
```

In fractional form: 3/10 + 4/10 = 7/10

In decimal form:

```
 .3
 .4
 --
 .7
```

Add

```
22 hundredths
 5 hundredths
-------------
27 hundredths
```

In fractional form: 22/100 + 5/100 = 27/100

In decimal form:

```
 .22
 .05
 ---
 .27
```

If you can add dollars and cents, you can add decimal fractions.

Add twenty-five cents + thirteen cents + eight cents = 46 cents

```
 .25
 .13
 .08
 ---
 .46
```

Add \$3.05 + \$.78 + \$40.00 =

```
$ 3.05
$  .78
$40.00
------
$43.83
```


PROCEDURE FOR ADDING DECIMALS

1. Add in column form.
2. All decimal points MUST be arranged in a vertical line, one under the other.
3. Add the numbers as though they were whole numbers.
4. Place the decimal point, in the sum, directly below the decimal points in the problem.

Add: .03 + .5 + .125

```
.03            .030
.5             .500
.125   or as   .125
----           ----
.655           .655
```

Add: 5.05 + .003 + .7 + 3.1

```
5.05            5.050
 .003            .003
 .7              .700
3.1     or as   3.100
-----           -----
8.853           8.853
```

Do empty places disturb you?

Then fill them in with zeros.

Will they change the value of the number? Let's see.

REMEMBER: Any number multiplied by 1 equals the number.

3/10 x 1 = 3/10

REMEMBER: Any number divided by itself equals 1.

2/2 = 1, 10/10 = 1, 100/100 = 1

Therefore:

.3 = 3/10 3/10 x 1 = 3/10

3/10 x 10/10 = 30/100 = .30

.3 = .30

OBSERVATION: When a decimal fraction ENDS in a zero, it does not change the value of the decimal fraction. .3 is the same as .30 or .300 or .3000 and so on.

TIME LIMIT—8 Minutes

ADDITION OF DECIMALS

Just remember one simple rule. KEEP THE DECIMAL POINTS LINED UP.
The position of the decimal point in the sum should be in line with the others.

Arrange vertically and add:

0.267 + 2.309 + 9.0923 + 23.104 + 15 =

7.004 + 0.062 + 24 + 60.0005 + 105.1 =

129.3 + 23.57 + .36 + 344 + 74.07 + .063 =

$8.76	$7.25	$ .33	$6.78	$7.71
.32	4.14	5.83	.87	5.27
1.44	.96	4.27	4.22	.89

6.95	.24	5.28	.74	.865
6.68	.67	.55	8.45	.976
4.15	5.86	.85	.36	.245

1.358	79.61	.96	.5731	23.
.09	4.99	8.71	.9091	.76
9.949	4.65	29.39	.5818	.35

.0419	.0411	.0858	.00879	.00304
.0081	.0072	.0071	.00052	.00165
.0952	.0327	.0457	.00602	.00746
.0398	.0541	.0282	.00236	.00527
.0805	.0500	.0221	.00430	.00485

64.00	87.00	.56
.96	37.00	.48
20.00	.97	92.00
.08	.73	.88

SCORING

1 error = excellent
2 or 3 errors = good

SUBTRACTING DECIMALS

The procedure for subtracting decimal fractions is the same as for whole numbers.

The only thing you have to pay attention to is the placement of the decimal point in the remainder.

If you can subtract dollars and cents, you can subtract decimal fractions.

EXAMPLES: Subtract 13 cents from 25 cents.

```
 .25
 .13
 .12  =  12 cents
```

Subtract \$2.25 from \$5.28.

```
$5.28
 2.25
 3.03
```

The zero indicates that there are no tenths—the zero is used as a placeholder.

EXAMPLE: Subtract \$5.33 from \$5.43.

```
$5.43
 5.33
  .10
```

Is .10 the same as .1? Of course it is.

REMEMBER: When a decimal fraction ENDS in a zero, it does not change the value of the decimal fraction.

Subtract \$1.25 from \$5.

```
$5.             $5.00
 1.25   or as    1.25
$3.75           $3.75
```

Note how the problem became easier to solve by filling in the empty places with zeros.

PROCEDURE FOR SUBTRACTING DECIMALS

1. Arrange the subtrahend under the minuend so that the decimal points are one under the other.
2. Subtract in the same way as you would with whole numbers.
3. Place the decimal point, in the remainder, directly below the decimal points in the problem.
4. Check the answer by adding the remainder to the subtrahend to equal the minuend.

EXAMPLES:

Subtract .012 from .528.

```
 .528         PROOF:    .516
 .012                   .012
 .516                   .528
```

Subtract .125 from .35.

```
.35    or as  .350   PROOF:    .225
.125          .125             .125
.225          .225             .350
```

Note how the problem became easier to solve by filling in the empty place with a zero.

Subtract .05 from .302.

```
.302   or as  .302   PROOF:    .252
.05           .050             .050
.252          .252             .302
```

Subtract .328 from 3.

```
3.      or as  3.000  PROOF:   2.672
 .328           .328            .328
2.672          2.672           3.000
```

SUBTRACTION OF DECIMALS

Just remember one simple rule. KEEP THE DECIMAL POINTS LINED UP.
The position of the decimal point in the remainder should be in line with the others.

Arrange vertically and subtract:

$5.83 − .96 =

$4.14 − 1.87 =

$7.71 − 4.42 =

$8.76 − 3.05 =

$1.44 − .96 =

$8.05 − .97 =

6.68 −2.87	5.28 −.85	8.45 −2.74	9.95 −3.67	10.05 −7.30
1.358 −.09	3.9091 −.3008	7.096 −.704	79.6 −.395	8.71 −3.907
.0419 −.0081	.0858 −.0071	.00602 −.00552	.0043 −.00236	.003 −.00165
64.00 −.96	37.00 −2.88	92.00 −10.18	29.00 −.0073	16.00 −.0088
.000897 −.00003	.000072 −.0000327	.0000746 −.0000052	.0000398 −.0000085	

SCORING

1 error = excellent
2 or 3 errors = good

MULTIPLYING DECIMALS

Multiplying decimals is much like multiplying whole numbers.
The only difference will be the placement of the decimal point in the product.

A half-dollar may be written as	$.50
and 3 times a half-dollar =	$1.50

The decimal .50 may be written as .5

REMEMBER: End zeros do not increase or decrease the value of a decimal fraction.

The problem above in decimal form is:

3 x .5 = 1.5

A quarter may be written as	$.25
and 3 times a quarter =	$.75

The same problem in decimal form is

3 x .25 = .75

A half of a half-dollar is a	quarter
and 1/2 x .50 =	.25

The fraction 1/2 may be written as .5

The same problem in decimal form is

.5 x .5 = .25

The work that is to follow will be no harder than the money problems above.

PROCEDURE:

1. Multiply as you would whole numbers.
2. Count the number of decimal places in both factors.
3. Starting with the extreme right digit in the product, count off to the left the number of places determined in Step 2 and place the decimal point in the product.

EXAMPLES:

One-half of $3 = 3 x.5

1. 3 x 5 = 15
2. The number of decimal places in both factors equals one.
3. Count off the same number of places from the right in the product and place the decimal point.

3 x.5 = 15 → 1.5

One-half of 30 cents = .3 x.5

1. 3 x 5 = 15
2. The number of decimal places in both factors equals two.
3. Count off two places and place the decimal point.

.3 x.5 = 15 → .15

One-half of $35 = 35 x.5

1. 35 x 5 = 175
2. The number of decimal places in both factors equals one.
3. Count off one place and place the decimal point.

35 x.5 = 175 → 17.5

One-half of $3.50 = 3.5 x.5

1. 35 x 5 = 175
2. The number of decimal places in both factors equals two.
3. Count off two places and place the decimal point.

3.5 x.5 = 175 → 1.75

TIME LIMIT–None

MULTIPLYING DECIMALS

Multiply:

2 × .3	2 × .5	4 × .7	4 × .5	6 × .4	6 × .5	8 × .9	8 × .5
.2 × .6	.2 × .5	.4 × .8	.4 × .5	.6 × .3	.6 × .5	.8 × .7	.8 × .5
2.2 × .7	2.2 × .5	4.4 × .4	4.4 × .5	6.6 × .7	6.6 × .5	8.8 × .6	8.8 × .5
.22 × 4	.22 × 5	.44 × 6	.44 × 5	.66 × 6	.66 × 5	.88 × 3	.88 × 5
.22 × .9	.22 × .5	.44 × .3	.44 × .5	.66 × .8	.66 × .5	.88 × .8	.88 × .5

SCORING

This one is on the house—
just a warm-up session.

MULTIPLYING DECIMALS (cont.)

If the product contains the correct digits but the decimal point is in the wrong position, the answer is wrong.

In the multiplication of decimal fractions, there's no need to line up the decimal points as in addition and subtraction.

EXAMPLE:

```
                                        4.2
1. 42 x 5 = 210                        x.05
2. The number of decimal places
   in both factors equals three.
3. Count off three places and
   place the decimal point.             4.2       4.2
                                       x.05      x.05
                                        210      .210
```

NOTE:
The extreme right digit is not 1 but 0.
The end zero must be kept and counted in the placing of the decimal point.
However, once the decimal point has been properly placed the end zero may be eliminated. .210 = .21

EXAMPLE:

```
                                          .0012
                                          x.012
1. 12 x 12 = 144
2. The number of decimal places
   in both factors equals seven.
3. Count off seven places                 .0012      .0012
   and place the decimal point.            .012       .012
                                             24         24
                                            12         12
                                        0000144   .0000144
```

FINDING THE PRODUCT OF THREE OR MORE FACTORS

EXAMPLE: .2 x .3 x .4 =

1. 2 x 3 = 6
   6 x 4 = 24
2. The sum of the decimal places equals three.
3. Count three places to the left and place the decimal point. = .024

EXAMPLE: .05 x .2 x .006 =

1. 5 x 2 = 10
   10 x 6 = 60
2. The sum of the decimal places equals six.
3. Count six places to the left and place the decimal point. = .000060 = .00006

EXAMPLE: .012 x .00012 x .22 x .0031 =

1. 12 x 12 = 144
   144 x 22 = 3168
   3168 x 31 = 98208
2. The sum of the decimal places equals fourteen
3. Count twelve places to the left and place the decimal point. = .00000000098208

TIME LIMIT–14 Minutes

MULTIPLYING DECIMALS

Multiply:

3.6 .06	4.4 .08	6.2 .09	5.7 .05	7.8 .07
.036 .07	.044 .06	.062 .08	.057 .09	.078 .05
.0036 .011	.0044 .013	.0062 .014	.0057 .015	.0078 .025

.3 x .5 x .4 =

.4 x .5 x .6 =

.06 x .3 x .08 =

.015 x .08 x .012 =

.008 x .013 x .0033 x .46 =

.5 x .0044 x .018 x .006 =

.2 x .7 x .5 =

.9 x .2 x .6 =

.08 x .3 x .05 =

.023 x .05 x .014 =

.015 x .0042 x .022 x .7 =

.03 x .0021 x .45 x .007 =

SCORING

1 error = excellent
2 errors = good

DECIMALS–DIVISION

A typical everyday decimal problem is,
How many half-dollars can one get out of a dollar? Of course, two.

The half-dollar may be written as a fraction–1/2 or as a decimal .50 / .5

The division problem in fractional form is: $\frac{1}{.5}$

REMEMBER: Any number multiplied by 1 equals the number.

Therefore $\frac{1}{.5} \times 1 = \frac{1}{.5}$

REMEMBER: Any number divided by itself equals 1.

Therefore $\frac{1}{.5} \times 1 = \frac{1}{.5} \times \frac{10}{10} = \frac{10}{5} = 2$

This means that $\frac{1}{.5}$ has the same value as $\frac{10}{5}$

Which of the two problems would you prefer to work with–the one with the decimal, or the one with the whole numbers?

Well then, let's attempt to change all forthcoming decimal problems, wherever possible, into whole-number division problems.

DIVIDING A WHOLE NUMBER BY A DECIMAL

$\frac{1}{.5} = .5\overline{)1}$

To convert .5 into a whole number, move the decimal point one place to the right.
This is the same as multiplying .5 by 10.

$5.\overline{)1}$

In order not to change the value of the problem, we must move the decimal point the same number of places in the dividend as in the divisor–one place to the right. This is the same as multiplying 1 by 10.

$5.\overline{)10.}$

OBSERVATION: To accommodate the new position of the decimal point, we had to call on the services of the placeholder 0.

The problem $.5\overline{)1}$ will be solved as $5\overline{)10}$ with quotient 2

Another typical everyday decimal problem is,
How many nickels can one get out of a dollar? Of course, twenty.

A nickel is equal to 5 cents.
A cent is one hundredth part of a dollar.
A nickel equals 5 hundredths of a dollar.

A nickel in decimal form is .05

The problem is written: $\frac{1}{.05} =$

$\frac{1}{.05} \times 1 = \frac{1}{.05} \times \frac{100}{100} = \frac{100}{5} = 20$

THE DIVISION PROCESS

$.05\overline{)1}$

To convert .05 to a whole number move the decimal point two places to the right.
This is the same as multiplying .05 by 100.

$05.\overline{)1}$

Now move the decimal point in the dividend the same number of places.

$05.\overline{)100.}$

The problem $.05\overline{)1}$ will be solved as $5\overline{)100}$ with quotient 20

TIME LIMIT—4 Minutes

DIVIDING A WHOLE NUMBER BY A DECIMAL

$.2\overline{)1}$ $.4\overline{)1}$ $.3\overline{)6}$ $.2\overline{)6}$

$.3\overline{)9}$ $.2\overline{)4}$ $.4\overline{)8}$ $.2\overline{)8}$

$.02\overline{)1}$ $.04\overline{)1}$ $.03\overline{)6}$ $.02\overline{)6}$

$.02\overline{)4}$ $.03\overline{)9}$ $.02\overline{)8}$ $.04\overline{)8}$

$.2\overline{)10}$ $.4\overline{)10}$ $.3\overline{)12}$ $.2\overline{)22}$

$.3\overline{)36}$ $.5\overline{)25}$ $.2\overline{)22}$ $.4\overline{)28}$

$.02\overline{)10}$ $.04\overline{)10}$ $.03\overline{)15}$ $.02\overline{)18}$

$.02\overline{)20}$ $.03\overline{)42}$ $.05\overline{)50}$ $.04\overline{)44}$

SCORING

1 error = excellent
2 errors = good

DECIMALS–DIVISION (cont.)

Another typical everyday decimal problem is,
Divide 75 cents among 3 people.
The answer is 25 cents per person.

75 cents in decimal form is: .75

The division problem in fractional form is: $\frac{.75}{3}$

The division process in decimal form is: $3\overline{).75}$

In this problem the divisor is not a decimal but a whole number; therefore, the problem of repositioning the decimal point does not exist.

Well, if the decimal point is not to be moved in the divisor, then it need not be moved in the dividend either.

DIVIDING A DECIMAL BY A WHOLE NUMBER

$3\overline{).75}$

1. Place the decimal point in the quotient directly above the decimal point in the dividend, like this: $3\overline{).75}$ (with the point above)

2. Divide as you would with whole numbers, like this:

```
   .25
3 ).75
    6
    15
    15
     0
```

PROOF: .25 x 3 = .75

Another typical everyday decimal problem is,
Divide $7.50 among 5 people.
The answer is $1.50 per person.

The division problem in fractional form is: $\frac{7.50}{5}$

The division process in decimal form is written as: $5\overline{)7.5}$ (with the point above)

Here again we have a divisor that is not a decimal but a whole number, and once again the problem of repositioning the decimal point does not exist.

And, we repeat, if the decimal point is not to be moved in the divisor, it will not be moved in the dividend.

DIVIDING A MIXED DECIMAL BY A WHOLE NUMBER

$5\overline{)7.5}$

1. Place the decimal point in the quotient directly above the decimal point in the dividend, like this: $5\overline{)7.5}$ (with the point above)

2. Divide as you would with whole numbers, like this:

```
  1.5
5 )7.5
   5
   25
   25
    0
```

PROOF: 1.5 x 5 = 7.5

TIME LIMIT–5 Minutes

DIVIDING DECIMALS

$2\overline{)\,.6}$	$4\overline{)\,.8}$	$3\overline{)\,.9}$	$2\overline{)\,6.2}$	$4\overline{)\,8.4}$	$3\overline{)\,9.6}$	
	$3\overline{)\,.12}$	$4\overline{)\,.48}$	$3\overline{)\,.33}$	$3\overline{)\,1.2}$	$4\overline{)\,4.8}$	$3\overline{)\,3.3}$
$5\overline{)\,.625}$	$6\overline{)\,.732}$	$7\overline{)\,.784}$	$5\overline{)\,6.25}$	$6\overline{)\,7.32}$	$7\overline{)\,7.84}$	
	$4\overline{)\,.496}$	$3\overline{)\,.669}$	$2\overline{)\,.886}$	$4\overline{)\,4.96}$	$3\overline{)\,6.69}$	$2\overline{)\,8.86}$
$5\overline{)\,.5125}$	$6\overline{)\,.7344}$	$7\overline{)\,.9135}$	$5\overline{)\,51.25}$	$6\overline{)\,73.44}$	$7\overline{)\,91.35}$	
	$8\overline{)\,.80944}$	$3\overline{)\,.99018}$	$4\overline{)\,.92016}$	$8\overline{)\,809.44}$	$3\overline{)\,990.18}$	$4\overline{)\,920.16}$

SCORING

1 error = excellent
2 errors = good

DECIMALS—DIVISION (cont.)

Another typical everyday decimal problem is,
How many nickels can one get out of a quarter? The answer is five.

A quarter in decimal form is: .25

A nickel in decimal form is: .05

The division problem in fractional form is: $\frac{.25}{.05}$

What will it take to change .05 into a whole number?
Multiply .05 by 100 to get 5.
This is the same as moving the decimal point 2 places to the right.

In order not to change the value of the problem, the dividend must also be multiplied by 100. $.25 \times 100 = 25$
This is the same as moving the decimal point 2 places to the right.

$$\frac{.25}{.05} \times \frac{100}{100} = \frac{25}{5} = 5$$

DIVIDING A DECIMAL BY A DECIMAL

.05) .25

1. To convert .05 to a whole number, move the decimal point 2 places to the right. — 05.) .25
2. Now move the decimal point in the dividend the same number of places. — 05.) 25.
3. Place the decimal point in the quotient directly above the new position of the decimal point in the dividend. — 5) 25 (decimal point above)
4. Divide as you would with whole numbers.

```
    5.
5 ) 25
    25
     0
```

PROOF: 5 x 5 = 25

Another typical everyday decimal problem is,
How many quarters can one get out of \$1.25? The answer is five.

\$1.25 is decimal form is: 1.25
A quarter in decimal form is: .25

The division problem in fractional form is: $\frac{1.25}{.25}$

What will it take to change .25 into a whole number?
Move the decimal point 2 places to the right.
This is the same as multiplying .25 by 100, to get 25.

In order not to change the value of the problem, the numerator must also have its decimal point moved 2 places to the right.
This is the same as multiplying 1.25 by 100, to get 125.

$$\frac{1.25}{.25} \times \frac{100}{100} = \frac{125}{25} = 5$$

DIVIDING A MIXED DECIMAL BY A DECIMAL

.25) 1.25

1. To convert .25 to a whole number, move the decimal point 2 places to the right. — 25.) 1.25
2. Now move the decimal point in the dividend the same number of places. — 25) 1 25.
3. Place the decimal point in the quotient directly above the new position of the decimal point in the dividend. — 25.) 1 25. (decimal point above)
4. Divide as you would with whole numbers.

```
      5
25 ) 125
     125
       0
```

PROOF: 5 x 25 = 125

TIME LIMIT—8 Minutes

DIVIDING DECIMALS

$.3\overline{).6}$	$.2\overline{).8}$	$.3\overline{).9}$	$.3\overline{)6.3}$	$.2\overline{)8.4}$	$.3\overline{)9.6}$
$.4\overline{)12}$	$.6\overline{).48}$	$.2\overline{).33}$	$.4\overline{)1.2}$	$.6\overline{)4.8}$	$.3\overline{)3.3}$
$.5\overline{).725}$	$.6\overline{).864}$	$.7\overline{).854}$	$.5\overline{)7.25}$	$.6\overline{)8.64}$	$.7\overline{)8.54}$
$.4\overline{).624}$	$.3\overline{).831}$	$.2\overline{).752}$	$.4\overline{)6.24}$	$.3\overline{)8.31}$	$.2\overline{)7.52}$
$.5\overline{).6175}$	$.6\overline{).8454}$	$.7\overline{).7056}$	$.5\overline{)61.75}$	$.6\overline{)84.54}$	$.7\overline{)70.56}$
$.8\overline{).83272}$	$.3\overline{).63915}$	$.4\overline{).94648}$	$.8\overline{)832.72}$	$.3\overline{)639.15}$	$.4\overline{)946.48}$

SCORING

1 error = excellent
2 errors = good

DIVIDING A WHOLE NUMBER BY A MIXED DECIMAL.

How many \$2.50s can one get out of \$5? Two.

2.5) 5

1. To convert 2.5 to a whole number, move the decimal point 1 place to the right.

   2 5.) 5

2. Move the decimal point in the dividend the same number of places.
   To do so, you must use a placeholder.

   2 5.) 5 0.

3. Divide as you would with whole numbers.

```
     2
25 ) 50
     50
      0
```

PROOF: 2 x 25 = 50

DIVIDING A MIXED DECIMAL BY A MIXED DECIMAL

How many \$1.25s can one get out of \$3.75? Three.

1.25) 3.75

1. To convert 1.25 to a whole number, move the decimal point 2 places to the right.

   1 25.) 3.75

2. Move the decimal point in the dividend the same number of places.

   1 25.) 3 75.

3. Divide as you would with whole numbers.

```
      3
125 ) 375
      375
        0
```

PROOF: 3 x 125 = 375

DIVIDING A SMALL WHOLE NUMBER BY A LARGE WHOLE NUMBER

Divide \$5 among 25 people.
The answer is 20 cents per person.

25) 5

1. The divisor is a whole number. There's no need for moving the decimal point.

2. If the decimal point is not moved in the divisor, it will not be moved in the dividend.

3. Place the decimal point in the quotient directly above the decimal point in the dividend (for every whole number, it is located behind the last digit).

   25) 5.

4. 5 = 5.0
   5 = 5.00
   5 = 5.000

   Add as many placeholders as is necessary to perform the division—they will not change the value of the whole number.

   25) 5.0

5. Divide as you would with whole numbers.

```
     .2
25 ) 5.0
     5 0
       0
```

PROOF: .2 x 25 = 5

TIME LIMIT–25 Minutes

DECIMAL DIVISION

$1.2\overline{)6}$	$1.5\overline{)9}$	$1.6\overline{)9.6}$	$1.3\overline{)9.1}$	$8\overline{)4}$	$6\overline{)3}$
$1.6\overline{)96}$	$1.3\overline{)91}$	$2.4\overline{)6.72}$	$3.3\overline{)1.584}$	$24\overline{)18}$	$40\overline{)9}$
$2.4\overline{)288}$	$3.3\overline{)495}$	$4.5\overline{)283.5}$	$5.2\overline{)38.48}$	$325\overline{)26}$	$525\overline{)84}$
$4.5\overline{)1{,}665}$	$5.2\overline{)1{,}456}$	$6.6\overline{)12.342}$	$7.3\overline{)210.97}$	$4{,}625\overline{)222}$	$2{,}500\overline{)32}$
$6.6\overline{)14{,}784}$	$7.3\overline{)23{,}944}$	$8.1\overline{)186.705}$	$9.4\overline{)2{,}825.64}$	$12{,}500\overline{)8}$	$12{,}800\overline{)96}$
$8.1\overline{)325{,}215}$	$9.4\overline{)282{,}752}$	$25.2\overline{)705.6}$	$31.8\overline{)1{,}971.6}$	$575{,}000\overline{)46}$	$620{,}000\overline{)496}$

SCORING

1 error = excellent
2 errors = good

CONVERTING DECIMALS TO FRACTIONS

.5 is read as 5 TENTHS.

Write the fraction the same way as you read it. In other words, use the place-value name as the denominator of the fraction.

$$.5 = 5 \text{ TENTHS} = \frac{5}{10}$$

If the fraction can be reduced, reduce it.

$$\frac{5}{10} = \frac{1}{2}$$

.25 is read as 25 HUNDREDTHS.

$$25 \text{ HUNDREDTHS} = \frac{25}{100} = \frac{1}{4}$$

.125 is read as 125 THOUSANDTHS.

$$125 \text{ THOUSANDTHS} = \frac{125}{1{,}000} = \frac{1}{8}$$

.00625 is read as 625 HUNDRED THOUSANDTHS.

$$625 \text{ HUNDRED THOUSANDTHS} = \frac{625}{100{,}000} = \frac{25}{4{,}000} = \frac{1}{160}$$

CONVERTING FRACTIONS TO DECIMALS

$$\frac{3}{4} = 4\overline{)3}$$

1. Place the decimal point in the quotient.
2. Use as many placeholders as are necessary.
3. Divide as you would with whole numbers.

```
     .75
4 ) 3.00
    2 8
     20
     20
      0
```

$$\frac{5}{8} = 8\overline{)5}$$

```
     .625
8 ) 5.000
    4 8
      20
      16
       40
       40
        0
```

$$\frac{9}{5} = 5\overline{)9}$$

```
     1.8
5 ) 9.0
    5
    40
    40
     0
```

TIME LIMIT–12 Minutes

CONVERTING DECIMALS AND FRACTIONS

Convert these decimals to fractions and reduce to lowest terms.

.8 = ______	.6 = ______	.3 = ______
.12 = ______	.15 = ______	.35 = ______
.375 = ______	.168 = ______	.665 = ______
.3125 = ______	.0125 = ______	.5685 = ______
4.5 = ______	6.8 = ______	15.3 = ______
3.85 = ______	4.48 = ______	11.05 = ______
12.125 = ______	15.075 = ______	28.072 = ______

Convert these fractions to decimals.

1/2 = ______	3/4 = ______
4/10 = ______	5/20 = ______
7/100 = ______	15/125 = ______
3 1/4 = ______	2 1/2 = ______
12 9/10 = ______	5 1/8 = ______
29 62/100 = ______	45 18/25 = ______
125 1/2 = ______	42 185/1,000 = ______
4/5 = ______	23 5/16 = ______
6/25 = ______	18 15/20 = ______
13/200 = ______	212 1/4 = ______
568 1/8 = ______	

SCORING

1 error = excellent
2 errors = good

DECIMALS–DIVISION (cont.)

DIVIDING A DECIMAL BY A MIXED DECIMAL

The problem $\frac{75 \text{ cents}}{\$1.50}$ may be read as:

1. How many $1.50s can one get out of 75 cents?
   —which sounds rather odd to many people.

2. 75 cents is what part of $1.50?
   —which sounds like a reasonable question and easy to answer—1/2

1.5) .75

1. To convert 1.5 to a whole number move the decimal point 1 place to the right. — 1 5.) .75
2. Move the decimal point in the dividend the same number of places. — 1 5.) 7.5
3. Place the decimal point in the quotient directly above the decimal point in the dividend (new position). — 1 5.) 7.5
4. Divide as you would with whole numbers.

```
     .5
15 ) 7.5
     7 5
     ---
       0
```

PROOF: .5 x 15 = 7.5

LARGER PROBLEMS

.025) .000625

1. To convert the divisor to a whole number, move the decimal point 3 places to the right. — 025.) .000625
2. Move the decimal point in the dividend the same number of places. — 025.) 000.625
3. Place the decimal point in the quotient directly above the new position of the decimal point in the dividend. — 025.) 000.625
4. Divide as you would with whole numbers—
   WITH THIS EXCEPTION:

   How many 25s can one get out of 6? NONE.
   PLACE A ZERO ABOVE THE 6.

```
      .0
25 ) .625
```

5. Continue with the normal division process.

```
      .025
25 ) .625
       50
      ---
       125
       125
      ---
         0
```

PROOF: .025 x 25 = .625

TIME LIMIT–15 Minutes

DIVIDING DECIMALS

2.3) .92	3.4) .85	1.4) .98	2.4) .792	1.6) .96
3.7) .851	1.25) .975	3.18) .9063	2.38) .8092	1.68) .7896
.025) .0006	.014) .000224	.018) .00027	.036) .000468	
.0128) .0005504	.0324) .0012312	.562) .0023042	.0417) .0015846	
.01292) .00054264	.02305) .0006454	.03712) .00159616	.04414) .0101522	

SCORING

1 error = excellent
2 errors = good

LARGER PROBLEMS

.0044) .00001452

1. In the divisor, move the decimal point 4 places to the right.

   0044.) .00001452

2. In the dividend, move the decimal point the same number of places.

   0044.) 0000.1452

3. Place the decimal point in the quotient directly above the new position of the decimal point in the dividend.

   0044.) 0000.1452

4. Divide as you would with whole numbers–

5. How many 44s can one get out of 1? NONE.
   PLACE A ZERO ABOVE THE 1.

   .0
   44) .1452

6. How many 44s can one get out of 14? NONE.
   PLACE A ZERO ABOVE THE 4.

   .00
   44) .1452

7. Continue with the normal division process.

```
     .0033
44 ) .1452
      132
       132
       132
         0
```

PROOF: .0033 x 44 = .1452

NOTE: If we did not insert the placeholders, the quotient may have read as
.33 or .033
If we multiply .33 by 44, we will not get .1452
If we multiply .033 by 44, we will not get .1452

LARGER PROBLEMS

.00064) .2048

1. In the divisor, move the decimal point 5 places to the right.

   00064.) .2048

2. In the dividend, move the decimal point the same number of places (use a placeholder).

   00064.) 20480.

3. Place the decimal point in the quotient directly above the new position of the decimal point in the dividend.

   00064.) 20480.

4. Divide as you would with whole numbers–

```
     3  .
64 ) 20480.
     192
      128
```

5. How many 64s in 204? 3
   Write 3 above the 4.

6. How many 64s in 128? 2
   Write 2 above the 8.

```
     32  .
64 ) 20480
     192
      128
      128
        0
```

DON'T STOP NOW.
Observe the position of the digit 2 and the position of the decimal point.
An empty space exists between them. It must be filled in with a placeholder.

```
     320.
64 ) 20480
     192
      128
      128
       00
       00
        0
```

PROOF: 320 x 64 = 20,480

NOTE: If we did not insert the placeholder, the quotient may have read as 32

If we multiply 32 by 64 we will not get 20,480.

TIME LIMIT–16 Minutes

LARGER PROBLEMS IN DECIMAL DIVISION

.0095) .0000323	.0037) .00002257	.00046) .3082	.000125) .3125	
	.0021) .00000336	.0238) .0001428	.00053) .2544	.000285) .1197
.0014) .00000504	.0461) .00037341	.00027) .2295	.000342) .12654	
	.0064) .000001024	.0446) .00028098	.00092) .2576	.000417) .10008
.0016) .00001468	.0069) .00003174	.00087) .5394	.000628) .2198	

SCORING

1 error = excellent
2 errors = good

DECIMALS–DIVISION (cont.)

LARGER PROBLEMS

.000125) .55

1. In the divisor, move the decimal point 6 places to the right.

000125.) .55

2. In the dividend, move the decimal point 6 places to the right. (Use 4 zero placeholders.)

000125.) 550000.

3. Place the decimal point in the quotient directly above the new position of the decimal point in the dividend.

000125.) 550000.

4. Divide as you would with whole numbers–

5. How many 125s can one get out of 550? 4
Write 4 above the first 0.

```
      4   .
125 ) 550000.
      500
       500
```

6. How many 125s can one get out of 500? 4
Write 4 above the second 0.

```
      44  .
125 ) 550000
      500
       500
       500
         0
```

DON'T STOP NOW.

Observe the position of the last digit in the quotient and the position of the decimal point.
There are 2 empty places that must be filled in with zero placeholders.

```
      4400.
125 ) 550000
      500
       500
       500
         0
```

PROOF: 4,400 x 125 = 550,000

NOTE: If the placeholders were not used, the quotient would be read as 44 and 44 x 125 will not give you 550,000.

ONCE AGAIN

.29) .08845

1. In the divisor, move the decimal point 2 places to the right.

29.) .08845

2. In the dividend, move the decimal point 2 places to the right.

29.) 08.845

3. Place the decimal point in the quotient directly above the new position of the decimal point in the dividend.

29.) 08.845

4. Divide as you would with whole numbers.

5. How many 29s can one get out of 88? 3

```
     .3
29 ) 8.845
     8 7
       14
```

6. How many 29s can one get out of 10? 0
Write 0 as a placeholder above the 4.

```
     .30
29 ) 8.845
     8 7
       145
```

7. How many 29s can one get out of 145? 5

```
     .305
29 ) 8.845
     8 7
       145
       145
         0
```

PROOF: .305 x 29 = 8.845

NOTE: If the zero placeholder were not used, the quotient would be read as .35 and .35 x 29 will not give you 8.845.

TIME LIMIT—25 Minutes

LARGER PROBLEMS IN DECIMAL DIVISION

.0040625) .52   .0009375) .24   .29) .1943   .118) .5782

.023125) .37   .00921875) .59   .18) .0774   .239) .10038

.0065625) .42   .005625) .18   .37) .0888   .319) .8613

.000859375) .22   .009375) .3   .49) .1127   .427) .016653

.0078125) .54   .000128) .44   .93) .03534   .829) .038963

SCORING

1 error = excellent
2 errors = good

DECIMALS–DIVISION (cont.)

REPEATING DECIMALS

Our work with decimals cannot be called complete until we show you the existence of REPEATING DECIMALS.

EXAMPLE $\frac{2}{3} = 3\overline{)2.0000000}$ = .6666666

```
     .6666666
  3 ) 2.0000000
      1 8
        20
        18
         20
         18
          2
```

No matter now many zeros we place after the decimal point in the dividend, we always keep getting 6s in the quotient. We've got to stop someplace–where? After 1 decimal point, 2, 3?

Why not write the quotient as .6. . . (.6 followed by three dots) to indicate that the number repeats itself indefinitely.

ROUNDING OFF DECIMALS

There are times when it pays to be brief with decimal answers by expressing them to the nearest TENTH, HUNDREDTH, THOUSANDTH, and so on.

Suppose we wish to round off .666 to the nearest hundredth– we will write it as: .666 → .67

If we round off .333 to the nearest tenth– we will write it as: .333 → .3

RULE (same as for whole numbers):
When the unwanted digit is 5 or greater than 5, INCREASE the last of the wanted digits by 1.

When the unwanted digit is 4 or less, the last of the wanted digits will remain AS IS.

DIVISION BY 10

$\frac{1}{10} = 10\overline{)1}$

1. Place the decimal point in the quotient directly above the decimal point in the dividend.
2. To perform the division, use a placeholder in the dividend (it will not change its value).
3. Divide as you would with whole numbers.

```
       .1
  10 ) 1.0
       1 0
         0
```

Dividing by 10 is easy—all you do is move the decimal point in the numerator 1 place to the left.

$\frac{1}{10} = .1$

$\frac{5}{10} = .5$ $\frac{.5}{10} = .05$ $\frac{.05}{10} = .005$

DIVISION BY 100

$\frac{1}{100} = 100\overline{)1}$

Dividing by 100 is easy – move the decimal point in the numerator 2 places to the left.

```
        .01
  100 ) 1.00
        1 00
           0
```

$\frac{8}{100} = .08$ $\frac{.8}{100} = .008$ $\frac{.08}{100} = .0008$

DIVISION BY 1,000

$\frac{1}{1{,}000} = 1{,}000\overline{)1}$

Dividing by 1,000 is easy– move the decimal point in the numerator 3 places to the left.

```
          .001
  1,000 ) 1.000
```

$\frac{25}{1{,}000} = .025$ $\frac{2.5}{1{,}000} = .0025$

NOTE: Another way of looking at it is: Move the decimal point as many places to the left as there are zeros in the denominator.

DECIMALS

Figure the following repeating decimals:

$3\overline{)1}$ $6\overline{)5}$ $9\overline{)4}$ $11\overline{)7}$ $11\overline{)5}$

$1.2\overline{)7}$ $.9\overline{)7}$ $1.1\overline{)5}$ $.6\overline{)1}$ $1.2\overline{)5}$

Divide by 10, 100, 1,000, and so on:

3/10 = ______ 5/100 = ______ 7/1,000 = ______ 9/10,000 = ______ 4/100,000 = ______

6/1,000,000 = ______ 15/100 = ______ 23/1,000 = ______ 37/10,000 = ______ 63/100,000 = ______

45/1,000,000 = ______ 115/1,000 = ______ 234/10,000 = ______ 513/100,000 = ______ 718/1,000,000 = ______

5,275/10,000 = ______ 2,871/100,000 = ______ 8,329/1,000,000 = ______ 32,753/100,000 = ______

72,561/1,000,000 = ________ 419,236/1,000,000 = ________

Round off to the nearest tenth:

.62 = ____ .48 = ____ .413 = ____ .568 = ____ .2823 = ____ 2.43 = ____ 15.69 = ____

Round off to the nearest hundredth:

.624 = ____ .489 = ____ .2823 = ____ .4479 = ____ .83479 = ____ 2.437 = ____ 15.693 = ____

Round off to the nearest thousandth:

.6247 = ______ .4894 = ______ .83479 = ______ .56882 = ______ 2.4374 = ______ 15.6939 = ______

DECIMALS–MULTIPLE OPERATIONS

What it means is that you're going to be faced with problems that will contain combinations of addition, subtraction, multiplication and division of decimals. The one thing you must remember, before solving such problems, is the ORDER the solving process must take.

ORDER OF OPERATIONS:

1. Simplify the numerator and denominator.
   a. Do all multiplication and division first in the order in which they appear FROM LEFT TO RIGHT.
   b. Do all addition and subtraction in the order in which they appear FROM LEFT TO RIGHT.
2. Then, solve all multiplication and division operations.
3. Last, do all subtractions and additions.

EXAMPLE 1: $\frac{2.3 + .04}{.9} = 2.6$

1. Simplify the numerator.

$$\begin{array}{r} 2.3 \\ +.04 \\ \hline 2.34 \end{array}$$

2. Divide. $.9\overline{)2.34} = 9\overline{)23.4}$

$$\begin{array}{r} 2.6 \\ 9\overline{)23.4} \\ \underline{18} \\ 54 \\ \underline{54} \\ 0 \end{array}$$

EXAMPLE 2: $\frac{4.8 - 1.2}{.006} = 600$

1. Simplify the numerator.

$$\begin{array}{r} 4.8 \\ -1.2 \\ \hline 3.6 \end{array}$$

2. Divide. $.006\overline{)3.6} = 6\overline{)3600}$

$$\begin{array}{r} 600 \\ 6\overline{)3600} \\ \underline{36} \\ 00 \\ \underline{00} \\ 0 \end{array}$$

EXAMPLE 3: $\frac{.0014}{.33 + .02} = .004$

1. Simplify the denominator.

$$\begin{array}{r} .33 \\ +.02 \\ \hline .35 \end{array}$$

2. Divide. $.35\overline{).0014} = 35\overline{).140}$

$$\begin{array}{r} .004 \\ 35\overline{).140} \\ \underline{140} \\ 0 \end{array}$$

EXAMPLE 4: $\frac{3.2 \text{ x } .04}{.0002} = 640$

1. Simplify the numerator.

$$\begin{array}{r} 3.2 \\ \text{x}.04 \\ \hline .128 \end{array}$$

2. Divide. $.0002\overline{).128} = 2\overline{)1280}$

$$\begin{array}{r} 640 \\ 2\overline{)1280} \\ \underline{12} \\ 08 \\ \underline{8} \\ 0 \end{array}$$

DECIMALS–MULTIPLE OPERATIONS

$\dfrac{.7 + .052}{.008} =$ $\dfrac{.22 + .2}{.06} =$ $\dfrac{.81 + .009}{.9} =$ $\dfrac{.005 + .00052}{.046} =$

$\dfrac{1.1 - .015}{.775} =$ $\dfrac{.005 - .00005}{.22} =$ $\dfrac{.00032 - .0001}{.005} =$ $\dfrac{.8 - .005}{.000795} =$

$\dfrac{.681}{.0027 + .0003} =$ $\dfrac{.00765}{.039 + .011} =$ $\dfrac{4.68}{.0012 + .0006} =$ $\dfrac{.0000189}{.00062 + .00001} =$

$\dfrac{5.6 \times .8}{.0004} =$ $\dfrac{.073 \times .5}{.04} =$ $\dfrac{.0036 \times .0015}{.00009} =$ $\dfrac{.89 \times .0025}{.5} =$

SCORING

1 error = excellent
2 errors = good

EXAMPLE 5:

$$\frac{.042 + .08 \times .003}{.000006} = 7{,}040$$

1. Simplify the numerator.
   REMEMBER: Multiplication comes first.

$$\begin{array}{r} .003 \\ \times .08 \\ \hline .00024 \end{array}$$

2. Now, the addition:

$$\begin{array}{r} .00024 \\ +.042 \\ \hline .04224 \end{array}$$

3. Divide.

$$.000006\overline{).04224} = 6\overline{)42240.} = 7040$$

$$\begin{array}{r} 7040 \\ 6\overline{)42240.} \\ 42 \\ \hline 024 \\ 24 \\ \hline 0 \end{array}$$

EXAMPLE 6:

$$\frac{.073 \times .007}{.073 + .007} = .0063875$$

1. Simplify the numerator.

$$\begin{array}{r} .073 \\ \times .007 \\ \hline .000511 \end{array}$$

2. Simplify the denominator.

$$\begin{array}{r} .073 \\ +.007 \\ \hline .08 \end{array}$$

3. Divide.

$$.08\overline{).000511} = 8\overline{).0511}$$

$$\begin{array}{r} .0063875 \\ 8\overline{).0511} \\ 48 \\ \hline 31 \\ 24 \\ \hline 70 \\ 64 \\ \hline 60 \\ 56 \\ \hline 40 \\ 40 \\ \hline 0 \end{array}$$

EXAMPLE 7:

$$\frac{.006 + 6.6 \div 2.2 \times .008}{.00002} = 1{,}500$$

1. Simplify the numerator.
   REMEMBER: Multiplication or division comes first in the order in which they appear FROM LEFT TO RIGHT.

$$2.2\overline{)6.6} = \begin{array}{r} 3 \\ 22\overline{)66} \\ 66 \\ \hline \end{array}$$

2. Multiply.

$$\begin{array}{r} .008 \\ \times 3 \\ \hline .024 \end{array}$$

3. Add.

$$\begin{array}{r} .024 \\ +.006 \\ \hline .03 \end{array}$$

4. Divide.

$$.00002\overline{).03} = \begin{array}{r} 1500 \\ 2\overline{)3000} \\ 30 \\ \hline 0 \end{array}$$

EXAMPLE 8:

$$\frac{.0000159}{.035 \times .02 - .0034 \times .05} = .03$$

1. Simplify the denominator.
   REMEMBER: Multiplication or division first in the order in which they appear FROM LEFT TO RIGHT. THEN add and subtract.

$$\begin{array}{r} .035 \\ \times .02 \\ \hline .00070 \end{array} \qquad \begin{array}{r} .0034 \\ \times .05 \\ \hline .000170 \end{array}$$

2. Now subtract.

$$\begin{array}{r} .00070 \\ -.00017 \\ \hline .00053 \end{array}$$

3. Divide.

$$.00053\overline{).0000159} = \begin{array}{r} .03 \\ 53\overline{)1.59} \\ 1\,59 \\ \hline 0 \end{array}$$

DECIMALS–MULTIPLE OPERATIONS

$\dfrac{.038 + .09 \times .004}{.00014} =$

$\dfrac{5.5594 + .23 \times .7}{.0012} =$

$\dfrac{.762 + .8 \times .12}{.05} =$

$\dfrac{.0088 + .042 \times .038}{.00008}$

$\dfrac{.044 \times .006}{.044 + .006} =$

$\dfrac{.21 \times .15}{.21 + .15} =$

$\dfrac{.00075 \times .00025}{.00075 + .00025} =$

$\dfrac{.0018 \times .0002}{.0018 + .0002} =$

$\dfrac{.35 + 3.5 \div .005 \times .35}{.35} =$

$\dfrac{.000009 + .0015 \div .2 \times .07}{.4} =$

$\dfrac{.00017 + .8002 \times .3 \div .004}{.005} =$

$\dfrac{.144}{.0084 \times .006 - .0041 \times .012} =$

$\dfrac{2.133}{3.8 \times .9 - 7.2 \times .08} =$

$\dfrac{2.5914}{75.14 \div .34 - .34 \div .0017} =$

SCORING

1 error = excellent
2 errors = good

Add:

.08 + .53 + .76 + .027 + 1.3 =

.39 + 4.05 + .5 + 7. + 2.93 =

.00196	37.09
.05523	23.44
.43948	.62
.00062	8.81
	.96

Subtract:

.27	3.08	23.005	.006271
–.05	–.69	–6.3	–.005385

Multiply:

38	4.9	.314	18.3
x.08	x.0012	x4	x .038

Divide:

$14\overline{).0056}$ $3.6\overline{).00108}$ $7\overline{).0040166}$

$.00254\overline{).9652}$ $.56\overline{)190.4}$ $.000023\overline{)1.2581}$

Add:

839	130,410	.00392	.709	1 1/2	10 3/4
607	41,713	.152	.803	2 3/4	2 3/8
725	196,429	.00008	.097	5 7/8	2 5/6
417	2,109,315	3.0067	.063		
274		.2	3.05		
			2.		

Subtract:

92,079	809,405	.00625	13.09	5 7/8	10 3/8
–84,602	–73,009	–.00189	–2.0025	–2 3/4	–2 5/6

Multiply:

417	8093	3.25	.0035
x83	x706	x.189	x.0062

6 1/8 x 3 3/7 =

12 5/6 x 4 4/11 =

Divide:

65) 30,420

128) 487,305

.025) 625

15) .000225

12 ÷ 5 2/3 =

5 4/5 ÷ 2 1/2 =

RELATED PROBLEMS–DECIMALS

Many different things can be thought of in terms of decimals.

AREA is the product of LENGTH and WIDTH.

How many square feet in a piece of material with these dimensions: length, 2.7 feet; width, 1.8 feet? ________________

VOLUME is the product of LENGTH times WIDTH times HEIGHT.

How many cubic feet in a piece of material with these dimensions: length, 6.3 feet; width, 5.2 feet; height, 7.7 feet? ________________

TIME A job cutting pieces of wood took 12.8 hours. If each piece took .4 hour, how many pieces were cut? ________________

WEIGHT If a box of dried fruit weighs 1.3 pounds, what is the total weight of 15 boxes? ________________

MONEY What is .6 of 3.5 dollars? ________________

GROUPS An order was received for molding 5.9 feet long, tied 8 to a bundle. How many bundles can be made from 708 feet? ________________

If 1 pound of rice costs $1.60, find the cost of a package weighing .7 pounds. ________________

If 1 quart of vinegar costs 64 cents, find the cost of one that contains .75 quart. ________________

If 1 pound of steak costs $1.44, find the cost of one weighing 1.75 pounds. ________________

RELATED PROBLEMS–DECIMALS

Find the total length, in feet, of the piece of wood below.

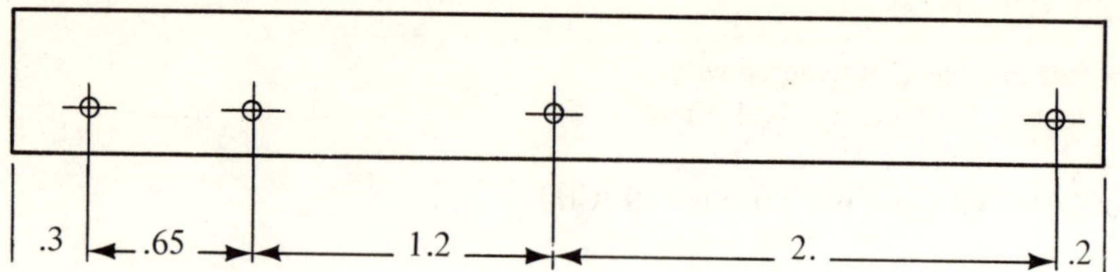

From a roll of ribbon 54.75 yards long, a piece 17.8 yards is cut.
How much ribbon is left on the roll? ____________

The rainfall for five days was as follows:

Monday	.2 inches
Tuesday	.13
Wednesday	.08
Thursday	1.1
Friday	.125

What was the total rainfall for the 5 days? ____________

Mother bought two pieces of roast beef that weighed 7.15 pounds together. If one piece weighed 3.8 pounds, what was the weight of the other piece? ____________

From a piece of lumber 12 feet long, two pieces were cut.
The first was 3.8 feet and the other was 6.9 feet.
How much of the original piece was used? ____________
How much of the original piece was left? ____________

A bottle was .8 full. If .45 of the contents were poured out, what part of the full bottle remained? ____________

A man can do a certain job in 1.2 hours.
A machine can do the same job in .2 of this time.
How long does it take the machine to do the job? ____________

Chapter 7

PERCENTS

PRETEST–PERCENTS

This area of arithmetic has a very high value as consumer education.

The ability of handling percents is needed in many financial transactions in the business world and in the daily life of the average citizen, who has to deal with:

Federal income tax returns
State and city sales taxes
Social security deductions and so on

A day doesn't go by when each individual is not involved with percents.

As in all prior chapters, we are offering you a pretest to check your knowledge of the subject.

What have you got to lose?

Take it, if only to find out your weak points.

We're sure that when you go through the chapter, you'll come out better equipped.

PRETEST–PERCENTS

Write as a percent:

.05 = ________ .28 = ________ 1.2 = ________ .002 = ________

Write in decimal form:

16% = ________ 3% = ________ .2% = ________ 125% = ________

Write as a fraction in lowest terms:

15% = ________ 37 1/2% = ________ 66 2/3% = ________ 25% = ________

6% of 120 = ________ 18 is 30% of ________ 48 is what percent of 320? ________

23% of 65 = ________ 24 is 37 1/2% of ________ 336 is what percent of 4,200? ________

1.5% of 42 = ________ 5 is 1% of ________ .08 is what percent of 16? ________

THE MEANING OF PERCENT

The number "fifteen hundredths" may be written

as a decimal .15

as a fraction $\frac{15}{100}$

The fraction 15/100 represents

15 parts out of 100 equal parts

Here is visual presentation of what we mean when we say

15 out of 100
or
15 per 100

We shall now study a third way of representing the number "fifteen hundredths."

This third way is called PERCENT.

PERCENT means "per hundred," "hundredths," "out of 100."

PERCENT is indicated by the symbol %.

15 percent may be written as	15%
15% means	15 hundredths
15 hundredths as a fraction is	15/100
15 hundredths as a decimal is	.15
15 hundredths as a percent is	15%

Obviously, fifteen hundredths may be written many ways:

as a fraction
as a decimal
as a percent

31 percent may be written as	31%
31% means	31 hundredths
31 hundredths as a fraction is	31/100
31 hundredths as a decimal is	.31
31 hundredths as a percent is	31%

45 percent may be written as	45%
45% means	45 hundredths
45 hundredths as a fraction is	45/100
45 hundredths as a decimal is	.45
45 hundredths as a percent is	45%

EXAMPLES:

Express .23 as a fraction:	23/100
Express 75% as a fraction:	75/100
Express 35% as a decimal:	.35
Express .67 as a percent:	67%
Express 55/100 as a decimal:	.55
Express 18/100 as a percent:	18%

ONCE AGAIN

Express .82 as a fraction:	82/100
Express 42% as a fraction:	42/100
Express 12% as a decimal:	.12
Express .33 as a percent:	33%
Express 9/100 as a decimal:	.09
Express 5/100 as a percent:	5%

PERCENTS

When we say 10 percent, we mean 10 parts for each ____________.

We can read 10 percent as ____________ hundredths.

Since percent means "hundredths," we can write 10 percent as a fraction having a denominator of ________.

The symbol ________ may be used instead of the denominator of 100.

The expression 22% is read as 22 ____________.

The expression 22% means ________ parts out of ____________.

The expression 22% may be read as ________ hundredths.

The expression 22% may be written as a fraction with a denominator of ________.

Since percent means "hundredths," write the following percents as fractions with denominators of 100.

2% ________ 12% ________ 26% ________ 31% ________ 43% ________

55% ________ 64% ________ 87% ________ 98% ________ 100% ________

How would you write the following fractions as percents?

3/100 ________ 11/100 ________ 23/100 ________ 35/100 ________

42/100 ________ 52/100 ________ 74/100 ________ 100/100 ________

THE MEANING OF PERCENT (cont.)

Let the figure represent 1 dollar
1 dollar = 100 cents
100 cents = 100 squares

A nickel is worth 5 cents.

5 cents will occupy 5 squares.

One nickel is 5 parts out of 100 equal parts.

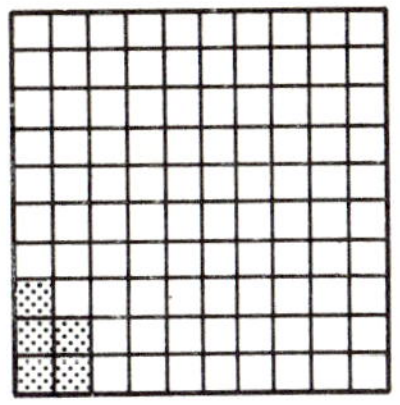

The relationship of one nickel to one dollar may be written:

As a fraction:	5/100 part of a dollar
As a decimal:	.05 part of a dollar
As a percent:	5% of a dollar

A quarter is worth 25 cents.

25 cents will occupy 25 squares.

One quarter is 25 parts out of 100 equal parts.

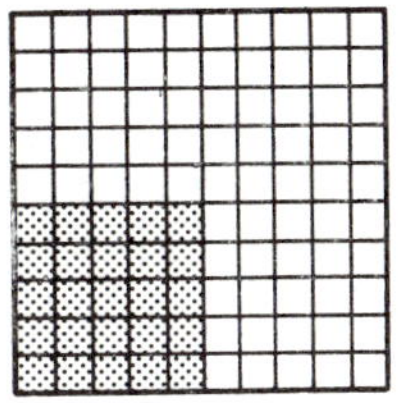

The relationship of one quarter to one dollar may be written:

As a fraction:	25/100 part of a dollar
As a decimal:	.25 part of a dollar
As a percent:	25% of a dollar

A half-dollar is worth 50 cents.

50 cents will occupy 50 squares.

One half-dollar is 50 parts out of 100 equal parts.

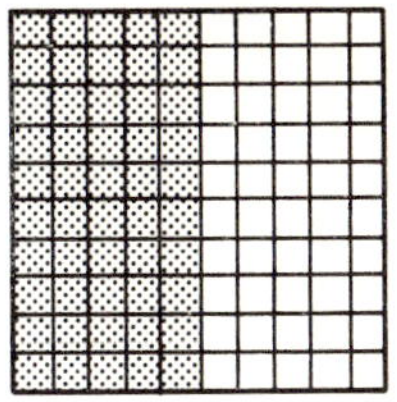

The relationship of one half-dollar to a dollar may be written:

As a fraction:	50/100
As a decimal:	.50 or .5
As a percent:	50%

One dollar is worth 100 cents.

100 cents will occupy 100 squares (the whole thing).

100 cents is one hundred hundredths of a dollar.

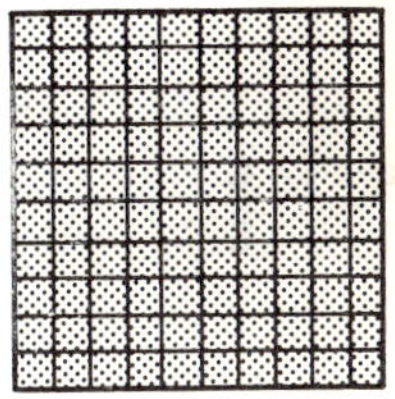

The relationship of 100 cents to a dollar may be written:

As a fraction:	100/100
As a decimal or whole number: (any number divided by itself = 1)	1
As a percent:	100%

OBSERVATION: PERCENT is not a new kind of number; it is merely a convenient way of expressing HUNDREDTHS.

PERCENT

50 cents is 50/100 or 1/2 of 100 cents.
50 cents is what percent of 100 cents? ________

25 cents is 25/100 or 1/4 of 100 cents.
25 cents is what percent of 100 cents? ________

10 cents is 10/100 or 1/10 of 100 cents.
10 cents is what percent of 100 cents? ________

5 cents is 5/100 or 1/20 of 100 cents.
5 cents is what percent of 100 cents? ________

1 cent is 1/100 of 100 cents.
1 cent is what percent of 100 cents? ________

25 cents is 25/50 or 1/2 of 50 cents.
25 cents is what percent of 50 cents? ________

10 cents is 10/20 or 1/2 of 20 cents.
10 cents is what percent of 20 cents? ________

5 cents is 5/10 or 1/2 of 10 cents.
5 cents is what percent of 10 cents? ________

1 cent is 1/2 of 2 cents.
1 cent is what percent of 2 cents? ________

5 dollars is what percent of 100 dollars? ________

5 dollars is what percent of 50 dollars? ________

5 dollars is what percent of 25 dollars? ________

5 dollars is what percent of 20 dollars? ________

5 dollars is what percent of 10 dollars? ________

5 dollars is what percent of 5 dollars? ________

The figures below contain 100 squares. Each square represents 1/100 of the figure. Therefore, each square must represent 1% of the figure.

Express in percents the shaded areas.

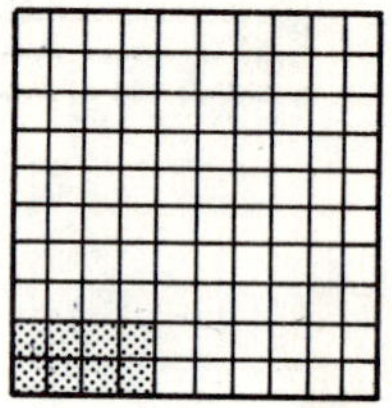

____

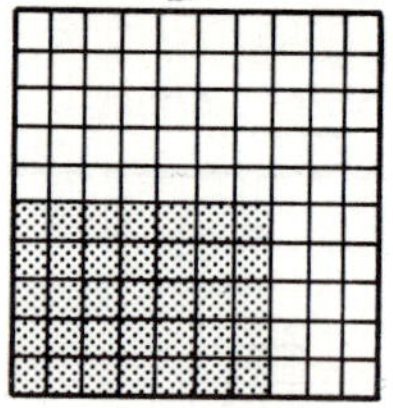

____

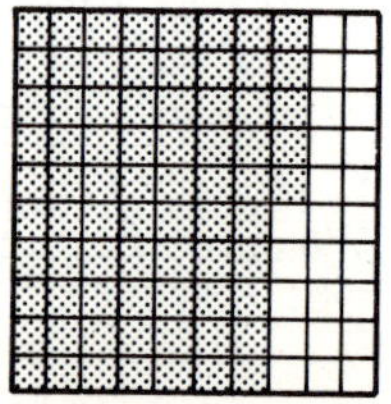

____

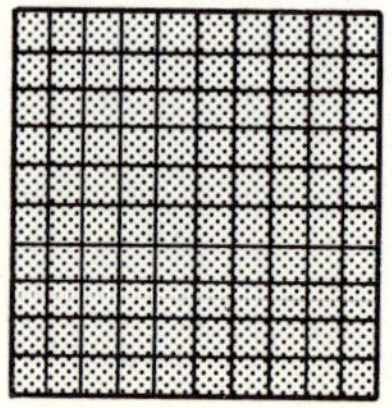

____

THE MEANING OF PERCENT (cont.)

Let the figure represent 1 dollar
1 dollar = 100 cents
100 cents = 100 squares

A nickel is worth 5 cents.

5 cents will occupy 5 squares.

One nickel is 5 parts out of 100 equal parts.

The relationship of one nickel to one dollar may be written:

As a fraction:	5/100 part of a dollar
As a decimal:	.05 part of a dollar
As a percent:	5% of a dollar

A half-dollar is worth 50 cents.

50 cents will occupy 50 squares.

One half-dollar is 50 parts out of 100 equal parts.

The relationship of one half-dollar to a dollar may be written:

As a fraction:	50/100
As a decimal:	.50 or .5
As a percent:	50%

A quarter is worth 25 cents.

25 cents will occupy 25 squares.

One quarter is 25 parts out of 100 equal parts.

The relationship of one quarter to one dollar may be written:

As a fraction:	25/100 part of a dollar
As a decimal:	.25 part of a dollar
As a percent:	25% of a dollar

One dollar is worth 100 cents.

100 cents will occupy 100 squares (the whole thing).

100 cents is one hundred hundredths of a dollar.

The relationship of 100 cents to a dollar may be written:

As a fraction:	100/100
As a decimal or whole number: (any number divided by itself = 1)	1
As a percent:	100%

OBSERVATION: PERCENT is not a new kind of number; it is merely a convenient way of expressing HUNDREDTHS.

TIME LIMIT–None

PERCENT

50 cents is 50/100 or 1/2 of 100 cents.
50 cents is what percent of 100 cents? ________

25 cents is 25/100 or 1/4 of 100 cents.
25 cents is what percent of 100 cents? ________

10 cents is 10/100 or 1/10 of 100 cents.
10 cents is what percent of 100 cents? ________

5 cents is 5/100 or 1/20 of 100 cents.
5 cents is what percent of 100 cents? ________

1 cent is 1/100 of 100 cents.
1 cent is what percent of 100 cents? ________

25 cents is 25/50 or 1/2 of 50 cents.
25 cents is what percent of 50 cents? ________

10 cents is 10/20 or 1/2 of 20 cents.
10 cents is what percent of 20 cents? ________

5 cents is 5/10 or 1/2 of 10 cents.
5 cents is what percent of 10 cents? ________

1 cent is 1/2 of 2 cents.
1 cent is what percent of 2 cents? ________

5 dollars is what percent of 100 dollars? ________

5 dollars is what percent of 50 dollars? ________

5 dollars is what percent of 25 dollars? ________

5 dollars is what percent of 20 dollars? ________

5 dollars is what percent of 10 dollars? ________

5 dollars is what percent of 5 dollars? ________

The figures below contain 100 squares. Each square represents 1/100 of the figure. Therefore, each square must represent 1% of the figure.

Express in percents the shaded areas.

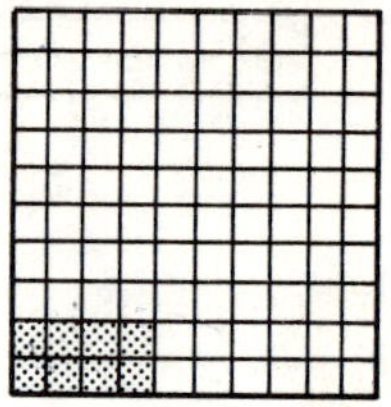

____

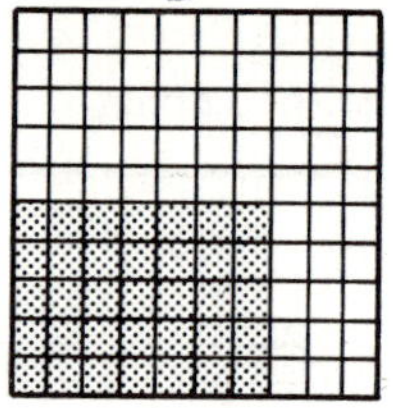

____

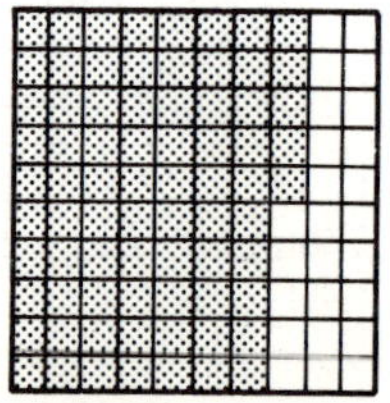

____

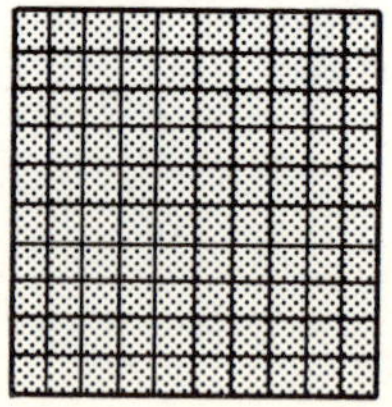

____

CHANGING PERCENTS TO DECIMALS OR MIXED DECIMALS

REMEMBER: Percent means "hundredths."
Hundredths means a fraction
with a denominator of 100. $\frac{\quad}{100}$

Hundredths is also the name
of the column 2 places
to the right of the decimal point.

If we can say that a digit located 2 places to the right of the decimal point is said to be in the hundredths column,
then,
we can say that the decimal point is located 2 places to the left of the hundredths column.

PROCEDURE for changing percents to decimals:

1. Drop the % sign.
2. Move the decimal point 2 places to the LEFT.

EXAMPLE: Change 15% to a decimal.

1. Drop the % sign. 15

2. Move the decimal point 2 places to the LEFT. .15

15% = .15

EXAMPLE: Change 5% to a decimal.

1. Drop the % sign. 5

2. Move the decimal point 2 places to the LEFT. .05 (Observe the use of a zero placeholder in the tenths column.)

5% = .05

EXAMPLE: Change 30% to a decimal.

1. Drop the % sign. 30

2. Move the decimal point 2 places to the LEFT. .30 or .3

REMEMBER: In a decimal fraction all zeros following the last digit may be eliminated without changing the value of the decimal fraction—therefore

.30 is the same as .3

30% = .30 or .3

EXAMPLE: Change 125% to a decimal or a mixed decimal.

1. Drop the % sign. 125

2. Move the decimal point 2 places to the LEFT. 1.25

125% = 1.25

EXAMPLE: Change 1/2% to a decimal.

1. Drop the % sign. 1/2

2. Change the fraction to a decimal. .5

3. Move the decimal point 2 places to the LEFT. .005
(Once again, observe the use of the zero placeholders.)

1/2% = .005

BY THE WAY, did you know that 1/2% may be written as .5%?

TIME LIMIT—4 Minutes

PERCENTS

Change the following percents to decimal fractions.

5% ________	2% ________	1% ________	10% ________	50% ________
25% ________	15% ________	75% ________	3.75% ________	1.75% ________
5.35% ________	8.45% ________	100% ________	250% ________	715% ________
332% ________	37 1/2% ________	83 1/3% ________	87 1/5% ________	15 1/8% ________
23% ________	66% ________	3% ________	18% ________	27% ________
16.4% ________	25.7% ________	46.2% ________	12.3% ________	33.8% ________
.47% ________	.63% ________	.15% ________	.98% ________	.1% ________
1/2% ________	1/4% ________	1/5% ________	1/8% ________	1/10% ________

SCORING

1 error = excellent
2 errors = good

CHANGING DECIMALS AND WHOLE NUMBERS TO PERCENTS

You learned to change a percent to a decimal by
1. dropping the % sign
2. moving the decimal point 2 places to the LEFT

like this: 48% = .48

Therefore, it seems reasonable that if we want to change a decimal to a percent, we will have to reverse the procedure.

PROCEDURE for changing decimals and whole numbers to percents:

1. Move the decimal point 2 places to the RIGHT.
2. Write the % sign behind the number.

EXAMPLE: Change .48 to a percent.

1. Move the decimal point 2 places to the RIGHT. 48.

2. Write the % sign behind the number. 48%

.48 = 48%

EXAMPLE: Change .02 to a percent.

1. Move the decimal point 2 places to the RIGHT. 02.

(Zeros in front of a whole number may be removed without changing the value of the number.) 2

2. Write the % sign behind the number. 2%

.02 = 2%

EXAMPLE: Change .035 to a percent.

1. Move the decimal point 2 places to the RIGHT. 3.5

2. Write the % sign behind the number. 3.5%

.035 = 3.5%

EXAMPLE: Change .002 to a percent.

1. Move the decimal point 2 places to the RIGHT. .2

2. Write the % sign behind the number. .2%

You read it as "two-tenths of 1 percent."
It may also be written as: 2/10%
or in its more simplified form: 1/5%
which is read as
"one-fifth of 1 percent"
.002 = .2% or 1/5%

EXAMPLE: Change 2 to a percent.

1. Move the decimal point 2 places to the RIGHT. 200.

REMEMBER: For a whole number the decimal point is understood to be located behind the last digit, like this: 2 = 2.

2. Write the % sign behind the number. 200%

2 = 200%

OTHER EXAMPLES:

.255	=	25.5%
.2	=	20%
1.2	=	120%
15.	=	1,500%

TIME LIMIT–4 Minutes

PERCENTS

Express the following whole numbers and decimal fractions as percents:

.05 ______	.10 ______	.01 ______	.5 ______	.025 ______
.25 ______	.005 ______	.205 ______	.385 ______	.015 ______
.735 ______	.008 ______	2.0 ______	5.0 ______	2.5 ______
1.25 ______	3.05 ______	1.65 ______	4.42 ______	6.255 ______
.0335 ______	.0862 ______	.0545 ______	.0095 ______	.6275 ______
.3345 ______	.2262 ______	.7648 ______	.125 ______	.375 ______
.625 ______	.875 ______	.0023 ______	.0076 ______	.0014 ______
.0099 ______	1.0525 ______	4.0036 ______	2.0237 ______	8.0725 ______

SCORING

1 error = excellent
2 errors = good

CHANGING FRACTIONS TO PERCENTS

REMEMBER: Percent means "hundredths." Hundredths may be written as a fraction with a denominator of 100.

IMPORTANT: When we want to change a fraction to a percent, we think only of a special fraction whose denominator is 100 and nothing else.

PROCEDURE for changing fractions to percents:

1. Remove the denominator of 100.
2. Write the % sign behind the number that was the numerator of the fraction.

EXAMPLE: Change 15/100 to a percent.

1. Remove the denominator. 15

2. Write the % sign behind the number. 15%

How simple can you make it?

15/100 = 15%

EXAMPLES:

5/100 = 5%

33/100 = 33%

100/100 = 100%

1,200/100 = 1,200%

25.5/100 = 25.5%

What about other fractions whose denominators are not 100, but are closely related, such as 10, 1,000, 10,000, and so on?

To solve such problems, all we need is to:

REMEMBER:
Any number multiplied by 1 equals the number.
Any fraction multiplied by 1 equals the number.
Any number divided by 1 equals the number.
Any fraction divided by 1 equals the number.

And the number 1 may be written in many forms;

such as 10/10, 100/100, 1,000/1,000.

Now, let's use this knowledge for the following.

EXAMPLE: Change 375/1,000 to a percent.

What will it take to change the denominator of 1,000 to a denominator of 100?

You are right—DIVIDE it by 10.

But if we divide the denominator by 10, we must also divide the numerator by 10 because 10/10 is only another symbol for the number 1—and any number multiplied or divided by 1 does not change the value of the number.

Therefore $\frac{375 \div 10}{1000 \div 10} = \frac{37.5}{100}$

1. Remove the denominator. 37.5

2. Write the % sign behind the number. 37.5%

375/1,000 = 37.5%

EXAMPLE: Change 7/10 to a percent.

$$\frac{7}{10} \times \frac{10}{10} = \frac{70}{100} = 70\%$$

TIME LIMIT—2 Minutes

PERCENT

Convert the following fractions to percents.

$\frac{2}{10}$ _____	$\frac{5}{10}$ _____	$\frac{9}{10}$ _____	$\frac{10}{10}$ _____	$\frac{20}{10}$ _____
$\frac{4}{10}$ _____	$\frac{7}{10}$ _____	$\frac{3}{10}$ _____	$\frac{15}{10}$ _____	$\frac{30}{10}$ _____
$\frac{2}{100}$ _____	$\frac{15}{100}$ _____	$\frac{70}{100}$ _____	$\frac{150}{100}$ _____	$\frac{300}{100}$ _____
$\frac{7}{100}$ _____	$\frac{5}{100}$ _____	$\frac{25}{100}$ _____	$\frac{40}{100}$ _____	$\frac{100}{100}$ _____
$\frac{22}{1,000}$ _____	$\frac{15}{1,000}$ _____	$\frac{70}{1,000}$ _____	$\frac{150}{1,000}$ _____	$\frac{300}{1,000}$ _____
$\frac{5}{1,000}$ _____	$\frac{25}{1,000}$ _____	$\frac{250}{1,000}$ _____	$\frac{700}{1,000}$ _____	$\frac{1,000}{1,000}$ _____

SCORING

1 error = excellent
2 errors = good

CHANGING FRACTIONS TO PERCENT (cont.)

When we use fractional notation, we must consider the existence of fractions with denominators of any number from 1 to infinity—and not only multiples of 10.

Within the group of possible denominators, we will find some that can easily be raised or lowered to 100, such as:

2	x	50	=	100	150	÷	1 1/2	= 100
4	x	25	=	100	200	÷	2	= 100
5	x	20	=	100	250	÷	2 1/2	= 100
8	x	12 1/2	=	100	300	÷	3	= 100
12	x	8 1/2	=	100	350	÷	3 1/2	= 100
20	x	5	=	100	400	÷	4	= 100
25	x	5	=	100	450	÷	4 1/2	= 100
50	x	2	=	100	500	÷	5	= 100

EXAMPLE: Change 3/4 to a percent.

We can change the denominator of 4 to 100 by multiplying it by 25, which of course means that we are obliged to do the same to the numerator.

Therefore $\frac{3}{4} \times \frac{25}{25} = \frac{75}{100}$

75/100 = 75%

EXAMPLE: Change 3/25 to a percent.

$\frac{3}{25} \times \frac{4}{4} = \frac{12}{100}$

12/100 = 12%

OTHER EXAMPLES:

2/5 = 40/100 = 40%

13/20 = 65/100 = 65%

17/50 = 34/100 = 34%

What are we going to do with the fraction 3/7?

Does a number exist that we can multiply by 7 to get a denominator of 100? NO.

With fractions of this type it is best to:

1. Change the fraction to a decimal.
2. Change the decimal to a percent.

EXAMPLE: Change 3/7 to a percent.

1. Divide the numerator by the denominator.

```
     .428
 7 ) 3.000
     2 8
       20
       14
        60
        56
         4
```

NOTE: 3 decimal places are sufficient.

The example left us with a remainder of 4. Since 4 is more than 1/2 of 7, we shall increase the value of the last digit by 1. 3/7 = .429

2. Now, to get the percent, move the decimal point 2 places to the right. 42.9

3. Write the % sign behind it. 42.9%

3/7 = 42.9%

EXAMPLE: Change 8/13 to a percent.

```
      .615
 13 ) 8.000
      7 8
        20
        13
         70
         65
```

This example left us with a remainder of 5. Since 5 is less than 1/2 of 13, we shall not increase the value of the last digit.

8/13 = .615 = 61.5%

TIME LIMIT–12 Minutes

PERCENTS

Change the following fractions to percents. One decimal place in the percent is sufficient.

$\frac{1}{2}$ ______	$\frac{1}{3}$ ______	$\frac{1}{4}$ ______	$\frac{1}{5}$ ______	$\frac{1}{6}$ ______
$\frac{1}{7}$ ______	$\frac{1}{8}$ ______	$\frac{1}{9}$ ______	$\frac{2}{3}$ ______	$\frac{3}{4}$ ______
$\frac{3}{5}$ ______	$\frac{3}{7}$ ______	$\frac{4}{5}$ ______	$\frac{4}{7}$ ______	$\frac{4}{9}$ ______
$\frac{3}{8}$ ______	$\frac{5}{6}$ ______	$\frac{5}{7}$ ______	$\frac{5}{8}$ ______	$\frac{6}{7}$ ______
$\frac{3}{16}$ ______	$\frac{3}{32}$ ______	$\frac{3}{64}$ ______	$\frac{5}{16}$ ______	$\frac{5}{32}$ ______
$\frac{18}{25}$ ______	$\frac{17}{20}$ ______	$\frac{19}{50}$ ______	$\frac{17}{30}$ ______	$\frac{23}{47}$ ______

SCORING

1 error = excellent
2 errors = good

CHANGING PERCENTS TO FRACTIONS OR MIXED NUMBERS

REMEMBER: Percent means "hundredths." Hundredths may be written as a fraction with a denominator of 100.

EXAMPLE: Change 5% to a fraction.

1. Use the numeral to represent the numerator. $\frac{5}{}$

2. Remove the % sign and in its place use the number 100 to represent the denominator. $\frac{5}{100}$

3. Reduce to lowest terms.

$$\frac{5}{100} \div \frac{5}{5} = \frac{1}{20} \qquad 5\% = 1/20$$

EXAMPLE: Change 37 1/2% to a fraction.

1. Use the numeral to represent the numerator. $\frac{37\ 1/2}{}$

   We know that 37 1/2 may also be written as $\frac{37.5}{}$

2. Remove the % sign and use 100 for the denominator. $\frac{37\ 1/2}{100}$

3. Reduce to lowest terms.

$$\frac{37\ 1/2}{100} \div \frac{12\ 1/2}{12\ 1/2} = \frac{3}{8}$$

37 1/2% = 3/8

EXAMPLE: Change 27% to a mixed number.

The reason we ask for a mixed number is that any percent 100% or over will have a value of 1 or more.

Do you want more proof?

1. Use the numeral to represent the numerator. $\frac{275}{}$

2. Remove the % sign and in its place use 100 as the denominator. $\frac{275}{100}$

We see a numerator that is greater than the denominator. Obviously the answer must be greater than 1. 2.75

275% = 2.75 = 2 3/4

OTHER EXAMPLES:

6%	=	6/100	=	3/50
10%	=	10/100	=	1/10
15%	=	15/100	=	3/20
25%	=	25/100	=	1/4
50%	=	50/100	=	1/2
33 1/3%	=	$\frac{33\ 1/3}{100}$	=	1/3
75%	=	75/100	=	3/4
100%	=	100/100	=	1
125%	=	125/100	=	1 1/4

TIME LIMIT–8 Minutes

PERCENTS

Change the following percents to whole numbers, mixed numbers, and fractions in their lowest terms.

1% =	2% =	3% =	4% =	5% =
6% =	7% =	8% =	9% =	10% =
15% =	20% =	25% =	30% =	35% =
40% =	45% =	50% =	55% =	60% =
65% =	70% =	75% =	80% =	85% =
90% =	95% =	100% =	125% =	150% =
175% =	200% =	250% =	300% =	500% =
12.5% =	37.5% =	62.5% =	87.5% =	33 1/3% =
66 2/3% =	.5% =	.2% =	.1% =	.025% =

SCORING

1 error = excellent
2 errors = good

OPERATIONS WITH PERCENTS

FINDING A PERCENT OF A NUMBER

Now that we have become familiar with how to change fractions and decimals to percents and percents to decimals and fractions,

let us concern ourselves with their use.

PERCENT is convenient to write or say, but it must first be converted to a fraction or a decimal in order to use it.

When we write a number such as 15%, it has no meaning at all by itself.

To have meaning, it must be 15% of something.

EXAMPLE: You are planning to buy a house. The cost of the house is $30,000. You are advised by the bank that a down payment of 15% is required to take over the house.

How much is 15% of $30,000?

1. Change 15% to a decimal. .15

2. Then multiply.

```
  $30,000
     x.15
   150000
   30000
$4,500.00
```

REMEMBER: There are a total of 2 decimal places in both factors. Therefore, we must move the D.P. (decimal point) 2 places to the left in our answer.

How much is 15% of $30,000? $4,500

EXAMPLE: Your gross income for 1972 is $12,500. The federal income tax is 22%. How much does Uncle Sam expect from you?

How much is 22% of $12,500?

1. Change 22% to a decimal. .22

2. Multiply.

```
  $12,500
     x.22
    25000
   25000
$2,750.00
```

22% of $12,500 is $2,750

EXAMPLE: You bought a suit for $87.50. The city sales tax is 7 1/2%. What will be the amount of the bill?

How much is 7 1/2% of $87.50?

1. Change 7 1/2% to a decimal. .075

2. Multiply.

```
  $87.50
   x.075
   43750
  61250
$6.56250
```

There are a total of 5 decimal places in both factors–move the D.P. 5 places to the left.

Tax = $6.56

The bill will be

```
 $87.50 suit
+  6.56 tax
 $94.06
```

TIME LIMIT–6 Minutes

PERCENTS

100% of 56 is ______

50% of 56 is ______

25% of 56 is ______

10% of 56 is ______

5% of 56 is ______

1% of 56 is ______

.1% of 56 is ______

100% of 128 is ______

50% of 128 is ______

25% of 128 is ______

10% of 128 is ______

5% of 128 is ______

1% of 128 is ______

.1% of 128 is ______

100% of 6,240 is ______

50% of 6,240 is ______

25% of 6,240 is ______

10% of 6,240 is ______

5% of 6,240 is ______

1% of 6,240 is ______

.1% of 6,240 is ______

What is 3% of 24? ______

What is 42% of 63? ______

What is 1% of 125? ______

What is 10% of 320? ______

What is 150% of 240? ______

What is 15% of 68? ______

What is .1% of 330? ______

What is 5% of 42? ______

What is 2% of 112? ______

What is 20% of 40? ______

What is 25% of 180? ______

What is 225% of 162? ______

What is 34% of 94? ______

What is .2% of 2,750? ______

SCORING

1 error = excellent
2 errors = good

OPERATIONS WITH PERCENTS

FINDING WHAT PERCENT ONE NUMBER IS OF ANOTHER

You bought an object for \$125.
You were given a discount of \$25.
The discount was equal to what percent?

EXAMPLE: \$25 is what percent of \$125?

Obviously, the answer has got to be less than 100% because:

25 is less than 125.

The problem may be reworded to read:

What percent of \$125 is \$25?

PROCEDURE: DIVIDE the number after the word "is" by the number after the word "of."

Think of it as: $\frac{\text{is}}{\text{of}} =$

1. Divide.

$$\begin{array}{r} .2 \\ 125\,\overline{)\,25.0} \\ \underline{25\;0} \\ 0 \end{array}$$

2. Convert the decimal to a percent.

.2 = 20%

PROOF: 20% of 125 =

$$\begin{array}{r} 125 \\ \underline{.2} \\ 25.0 \end{array}$$

EXAMPLE: Suppose the problem read:
\$125 is what percent of \$25?

Reworded: What percent of \$25 is \$125?

REMEMBER: $\frac{\text{is}}{\text{of}}$

1. Divide.

$$\begin{array}{r} 5 \\ 25\,\overline{)\,125} \\ \underline{125} \\ 0 \end{array}$$

2. Convert the whole number to a percent.

5 = 500%

PROOF: 500% of \$25 is

$$\begin{array}{r} 25 \\ \underline{\text{x}5} \\ 125 \end{array}$$

EXAMPLE: Johnny took a test containing 40 problems.
He worked 25 problems correctly.
What percent of the problems did he work out correctly?

25 is what percent of 40?

Reworded: What percent of 40 is 25?

1. Divide. $\frac{\text{is}}{\text{of}}$

$$\begin{array}{r} .625 \\ 40\,\overline{)\,25.000} \\ \underline{240} \\ 100 \\ \underline{80} \\ 200 \\ \underline{200} \\ 0 \end{array}$$

2. Convert the decimal to a percent. .625 = 62.5%

PROOF: 62.5% of 40 = 25

PERCENTS

640 is what percent of 64? ____________

64 is what percent of 64? ____________

32 is what percent of 64? ____________

16 is what percent of 64? ____________

6.4 is what percent of 64? ____________

3.2 is what percent of 64? ____________

.64 is what percent of 64? ____________

.064 is what percent of 64? ____________

15,000 is what percent of 1,500? ____________

1,500 is what percent of 1,500? ____________

750 is what percent of 1,500? ____________

375 is what percent of 1,500? ____________

150 is what percent of 1,500? ____________

75 is what percent of 1,500? ____________

15 is what percent of 1,500? ____________

1.5 is what percent of 1,500? ____________

What percent of 80 is 20? ____________

What percent of 150 is 15? ____________

What percent of 120 is 60? ____________

What percent of 600 is 6? ____________

What percent of 160 is 8? ____________

What percent of 36 is 12? ____________

What percent of 75 is 50? ____________

What percent of 90 is 90? ____________

What percent of 50 is 500? ____________

What percent of 240 is 30? ____________

200 is what percent of 1,000? ____________

12 is what percent of 120? ____________

24 is what percent of 48? ____________

1.75 is what percent of 35? ____________

.6 is what percent of 60? ____________

250 is what percent of 750? ____________

273 is what percent of 65? ____________

9.6 is what percent of 64? ____________

15.84 is what percent of 72? ____________

1.32 is what percent of 88? ____________

SCORING

1 error = excellent
2 errors = good

OPERATIONS WITH PERCENTS

FINDING A NUMBER WHEN ONLY A PERCENT OF IT IS KNOWN

EXAMPLE:

Mr. Jones is a salesman.
His commission is 8% based on the amount of his sales.
What total sales must he make to earn $200 a week?

8% of what number is 200?

Reworded: 200 is 8% of what number?

Obviously, the answer must be much larger than 200, and that we can only get if we divide the given number by the percent.

PROCEDURE:

1. Change the percent to a decimal.

8% = .08

2. Divide.

```
                    2500.
.08 ) 200.  =  8 ) 20000.
                   16
                    40
                    40
```

8% of what number is $200? = $2,500

PROOF: 2,500 x .08 = 200

HELPFUL HINT:

1% of 500	=	5
10% of 500	=	50
25% of 500	=	125
50% of 500	=	250
100% of 500	=	500
200% of 500	=	1,000
500% of 500	=	2,500

MENTAL EXERCISE

If 5 is 10% of a number, then the number must be 50

If 8 is 50% of a number, then the number must be 16

If 20 is 25% of a number, then the number must be 80

If 500 is 200% of a number, then the number must be 250

EXAMPLE: 210 is 35% of what number?

1. Change to a decimal.
2. Divide.

```
                     600
.35 ) 210  =  35 ) 21000.
                   210
                     0
```

PROOF: 35% of 600 = 210

EXAMPLE: 40 is 250% of what number?

1. Change to a decimal.
2. Divide.

```
                   16
2.5 ) 40  =  25 ) 400
                  25
                  150
                  150
                    0
```

PROOF: 250% of 16 = 40

EXAMPLE: 5 is .4% of what number?

1. Change to a decimal.
2. Divide.

```
                   1,250
.004 ) 5  =  4 ) 5,000
                 4
                 1 0
                   8
                   20
                   20
                    0
```

PROOF: .4% of 1,250 = 5

TIME LIMIT—10 Minutes

PERCENTS

200 is 1,000% of	______	1,000% of what number is 45?	______
200 is 100% of	______	100% of what number is 45?	______
200 is 50% of	______	50% of what number is 45?	______
200 is 25% of	______	25% of what number is 45?	______
200 is 10% of	______	10% of what number is 45?	______
200 is 5% of	______	5% of what number is 45?	______
200 is 1% of	______	1% of what number is 45?	______
200 is .1% of	______	.1% of what number is 45?	______

20 is 10% of	______	15 is 8% of	______
15 is 20% of	______	12 is 12 1/2% of	______
112 is 25% of	______	14 is 35% of	______
50 is 50% of	______	16 is 80% of	______
260 is 100% of	______	30 is 37 1/2% of	______
175 is 1,000% of	______	.75 is 2% of	______
175 is 5% of	______	1.63 is 1% of	______
6 is 1% of	______	15.25 is 5% of	______
1.5 is 2% of	______	.85 is .1% of	______
60 is 33 1/3% of	______	.075 is .05% of	______

SCORING

1 error = excellent
2 errors = good

TIME LIMIT—6 Minutes

ACHIEVEMENT TEST 7

Express as a decimal:

8% = ________ 15% = ________ 80% = ________ 225% = ________

Express as a percent:

.07 = ________ .55 = ________ .002 = ________ 3.5 = ________

Express as a fraction in lowest terms:

20% = ________ 87 1/2% = ________ 33 1/3% = ________ 75% = ________

Find the following:

23% of 48 = ________ 65% of 50 = ________

5 1/2% of 88 = ________ 150% of 142 = ________

20% of what number is 60? ________ 28 is 35% of what number? ________

45% of what number is 90? ________ 63 is 150% of what number? ________

30 is what percent of 120? ________ 48.6 is what percent of 324? ________

108 is what percent of 864? ________ 3.3 is what percent of 66? ________

TIME LIMIT–10 Minutes

REFRESHER TEST 6

Add:

44	47.08	8.294	27,402
23	12.76	.695	1,796
37	65.89	.482	38,347
98	9.26	5.769	4,745
	38.97	.789	783
		.037	15,486
		6.386	

Subtract:

726	80.2	93.19	530,192
487	27.9	48.74	365,788

Multiply:

4.5	806	3.48	1,097
3.5	29	20.8	6,802

Divide:

$8\overline{)5.3}$ $7\overline{).203}$ $7\overline{)34235}$

$19\overline{)1706}$ $.69\overline{)4899}$ $.076\overline{)521.512}$

RELATED PROBLEMS–PERCENTS

Mr. Brown works as a salesman. His commission is 7% of his sales. What commission will he receive in dollars and cents on a total sales of $4,500? __________

A company shipped 36,000 pieces. 3% of the pieces were returned as damaged. How many were returned? __________

A cash-checking establishment charges 1/2% for its service. What is the charge for cashing a $250 check? __________

The list price for a color TV set is $495. At a discount house you can get 20% off. What will it cost you at the discount house? __________

Chlordane is used for exterminating termites. The solution sold at the hardware stores is a 2% solution. This means that 2% of the full contents is pure chlordane; the rest is plain water. In a quart bottle that contains a full 32 ounces, what part is pure chlordane, and what part is water? __________ __________

In a hospital, dextrose is fed intravenously to patients. The solution is usually a 5% solution. This means 5% pure sugar and 95% plain water.
In a quart bottle, how many ounces are pure sugar? __________
How many ounces are plain water? __________

If the tax rate is 4.17 per 100 of assessed valuation, what is the amount of the tax on property valued at $55,000? __________

A baseball team played 52 games, which represents 33 1/3% of the total number of games it must play. How many games does the team have to play? __________

RELATED PROBLEMS–PERCENTS

In our town, the United Fund raised $6,300, which represented 28% of its quota. What is the town's quota? ____________

Mr. Jones made a down payment of $3,500 when he bought his new house. This was to be 10% of the price of the house. What was the price of the house? ____________

What percent of a quart is 12 fluid ounces? ____________
(32 fluid ounces = 1 quart)

What part of a pound is 10 ounces? ____________
(16 ounces = 1 pound)

If a baseball player is at bat 320 times and is able to get 102 hits, what is his batting average? ____________

A man bought a car for $3,200.
He had to pay a tax of $192.
What percent of the value of the car was the tax? ____________

In a class of 45 students, 30 passed a given test.
What percent of the class passed the test? ____________

During a baseball season, the Mets won 90 games and lost 78.
What percent of the games played did they win? (3 decimal places) ____________
What percent of the games played did they lose? ____________

In a semester, there are 60 school days.
If Billie was absent 8 days, what percent of the school days was she absent? ____________

GENERAL MATHEMATICS TEST

GENERAL MATHEMATICS TEST

TIME LIMIT—20 Minutes

Add:

$$\begin{array}{r} 8 \\ 7 \\ 6 \\ 5 \\ +9 \\ \hline \end{array}$$

Subtract:

$$\begin{array}{r} 1{,}244 \\ -\ 659 \\ \hline \end{array}$$

Add:

$$\begin{array}{r} 5{,}087 \\ +8{,}076 \\ \hline \end{array}$$

Subtract:

$$\begin{array}{r} 700 \\ -\ 353 \\ \hline \end{array}$$

Add:

$$\begin{array}{r} 96 \\ 78 \\ 69 \\ 83 \\ +88 \\ \hline \end{array}$$

Subtract:

$$\begin{array}{r} 3{,}645.38 \\ -\ 767.39 \\ \hline \end{array}$$

Add:

$$\begin{array}{r} 9{,}387 \\ 775 \\ +\ 566 \\ \hline \end{array}$$

Subtract:

$$\begin{array}{r} 1{,}392.05 \\ -\ 796.77 \\ \hline \end{array}$$

Add:

$$\begin{array}{r} 57 \\ 96 \\ 66 \\ 78 \\ +89 \\ \hline \end{array}$$

Subtract:

$$\begin{array}{r} 42{,}020 \\ -40{,}575 \\ \hline \end{array}$$

Add:

$$\begin{array}{r} 878 \\ 469 \\ 526 \\ 887 \\ +679 \\ \hline \end{array}$$

Subtract:

$$\begin{array}{r} 22{,}035 \\ -7{,}969 \\ \hline \end{array}$$

Multiply:

$$\begin{array}{r} 6{,}080 \\ \times\ 78 \\ \hline \end{array}$$

Divide:

$$66\,\overline{)\,5{,}808}$$

Multiply:

$$\begin{array}{r} 4{,}684 \\ \times\ 76 \\ \hline \end{array}$$

Divide:

$$63\,\overline{)\,31{,}941}$$

Multiply:

$$\begin{array}{r} 887 \\ \times\ 59 \\ \hline \end{array}$$

Divide:

$$54\,\overline{)\,48{,}816}$$

Multiply:

$$\begin{array}{r} 6{,}487 \\ \times\ 8 \\ \hline \end{array}$$

Divide:

$$45\,\overline{)\,3{,}888}$$

Multiply:

$$\begin{array}{r} 800 \\ \times\ 657 \\ \hline \end{array}$$

Divide:

$$34\,\overline{)\,2{,}528}$$

Multiply:

$$\begin{array}{r} 590 \\ \times\ 678 \\ \hline \end{array}$$

Divide:

$$68\,\overline{)\,43{,}554}$$

Multiply:

$$\begin{array}{r} 8{,}297 \\ \times\, 86 \\ \hline \end{array}$$

Divide:

$65\,\overline{)\,450.2}$

Add:

$$\frac{3}{4} + \frac{4}{7}$$

Subtract:

$$\frac{2}{3} - \frac{3}{5}$$

$3\ 3/5 \times 10 =$

$3\ 3/5 \div 3\ 3/10 =$

Add:

$$\frac{1}{6} + \frac{2}{3}$$

Subtract:

$$\begin{array}{r} 4\ 3/16 \\ -\ 3/4 \\ \hline \end{array}$$

Subtract:

$$\frac{7}{8} - \frac{1}{2}$$

$24 \times 5\ 3/8 =$

$2 \div 12 =$

Add:

$$\frac{1}{9} + \frac{5}{9}$$

Subtract:

$$\begin{array}{r} 4\ \ 2/3 \\ -1\ 11/12 \\ \hline \end{array}$$

$1/5 \times 1/5 =$

$4\ 2/7 \div 3 =$

Divide:

$6\,\overline{)\,.9}$

Convert to a decimal:

$\frac{4}{5}$ = ________

Multiply:

4.15
× 5
____

What is 175% of 200? ________

$8\overline{)1.04}$

Convert to a decimal:

3 1/4% = ________

$4.2\overline{)126}$

2 is what percent of 200? ________

$.6\overline{)3}$

What is 7 1/2 of $30? ________

Express .3 as a percent. ________

$100 is what percent of $4,000? ________

$3.5\overline{)1{,}422.4}$

$225 is 3 3/4% of what number? ________

$3.6\overline{)1{,}872}$

ANSWERS

ANSWER SHEET–Chapter 1

Page 4 Pretest

21 32 44 26 35 57 37 43

37 51

33 367 390 16,166 11,130 21,349 185,628,692 49,535

1,002 9,302

56 579 5,082 52,286

11

12

8

54

65

Page 6

1. idea
2. numeral
3. ten
4. digit
5. ten
6. ten
7. digits
8. order
9. number line
10. increase
11. right
12. 5
13. left
14. 7
15. numeral

Page 8

1.	3	9.	5
2.	22	10.	62
3.	130	11.	240
4.	213	12.	324
5.	333	13.	444
6.	404	14.	707
7.	555	15.	959
8.	300	16.	800

Page 10

1. ten
2. ten
3. ones
4. three
5. hundreds
6. value
7. position
8. place value
9. placeholder
10. nine
11. nine
12. nine
13. 6 hundreds
2 tens
4 ones
14. 2 thousands
0 hundreds
6 tens
3 ones
15. 35
16. 53
17. 5 tens
5 thousands
5 ones
5 ten thousands
5 hundred thousands
18. ten
19. ten
20. hundred

Page 12

1. 4 (10) + 5 (1)
2. 1 (100) + 7 (10) + 8 (1)
3. 3 (1,000) + 4(100) + 6 (10) + 7 (1)
4. 1 (10,000) + 5 (1,000) + 7 (100) + 0 (10) + 9 (1)
5. 7 (100,000) + 8 (10,000) + 3 (1,000) + 5 (100) + 5 (10) + 2 (1)

1. 54
2. 306
3. 5,270

1. 56,903
2. 788,304
3. 1,667,893
4. 56,637,721
5. 922,508,347
6. 1,005,004,006

1. Five thousand
2. Fifty-five thousand
3. Five hundred fifty-five thousand
4. Five thousand one hundred
5. Fifty-five thousand one hundred
6. Five hundred fifty-five thousand one hundred
7. Five thousand ten
8. Fifty-five thousand ten
9. Five hundred fifty-five thousand ten
10. Five thousand one hundred ten
11. Fifty-five thousand one hundred ten
12. Five hundred fifty-five thousand one hundred ten
13. Five million five hundred fifty-five thousand one hundred ten
14. Three billion six hundred thirty-six million four hundred twenty-two thousand one

1. 5,035
2. 87,108
3. 205,069
4. 6,006,006
5. 532,655,035

ANSWER SHEET–Chapter 1

Page 22

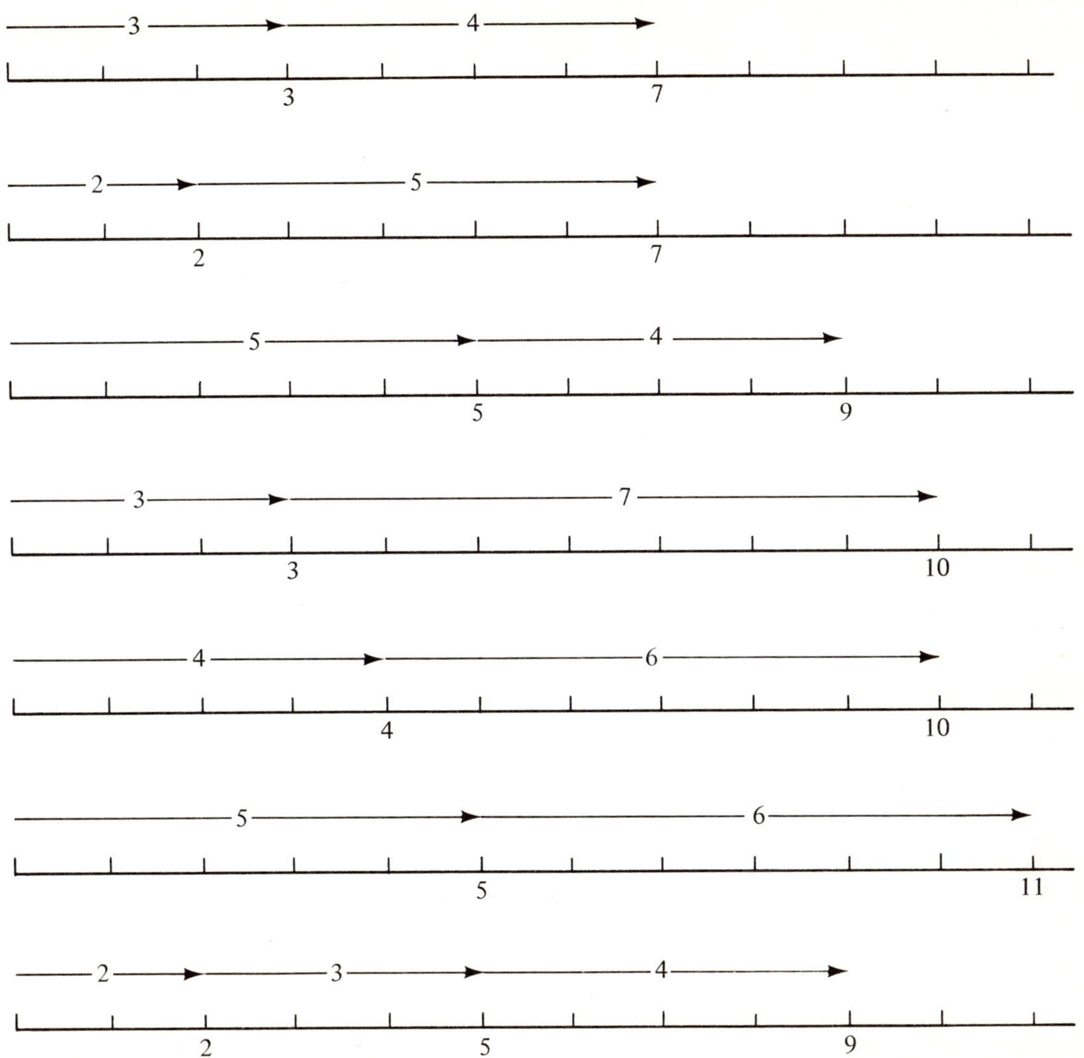

Page 24

6	7	10	8	7	6	2	9	8	11
11	9	7	10	3	7	10	12	9	8
4	11	10	13	13	7	12	10	11	11
7	8	13	12	9	14	5	11	12	14
15	11	9	15	14	9	15	11	13	13
13	10	12	15	11	12	9	16	15	7
11	12	14	15	9	17	8	11	15	10
16	13	11	14	12	9	14	15	17	11

Page 26

17	23	18	24	21	19	25	20	22
19	25	20	26	23	21	27	22	24
16	22	17	23	20	18	24	19	21
18	24	19	25	22	20	26	21	23
20	26	21	27	24	22	28	23	25

33	31	36	30	32	35	29	34	28
31	29	34	28	30	33	27	32	26
34	32	37	31	33	36	30	35	29

Page 26

17
19
18
20
22
20
24
25
29
30
32
36
40
44
47
50
61
71
78
89

Page 28

9	8
69	88
13	9
43	59
13	11
93	81

		63
		75
	13,	93
	15,	45
50,	14,	64
70,	12,	82

		42
		53
	13,	33
	13,	53
60,	11,	71
80,	17,	97

18	19	19	19	16	16	19	19
29	38	29	39	29	39	28	38
48	59	47	59	49	58	49	58
66	79	69	78	66	76	67	79
23	23	21	21	25	21	21	22
31	43	33	41	33	45	33	43
53	64	55	63	54	62	55	66
77	82	73	86	73	85	73	84

ANSWER SHEET–Chapter 1

Page 30

11, 23
12, 23

9, 23
14, 23

15, 28
13, 28

13, 25
12, 25

24, 30
21, 30

20, 28
20, 28

18, 25
19, 25

18, 27
19, 27

14	15	13	14	14	14	12
17	19	19	23	20	17	15
16	21	23	28	23	21	18
26	26	26	26	26	26	26
30	31	29	30	30	30	27

Page 32

13, 22, 26
13, 20, 26

13, 16, 24
11, 19, 24

16, 21, 30
14, 23, 30

14, 22, 28
14, 19, 28

15, 19, 27, 35, 42
15, 23, 27, 36, 42

25	22	24	23	25	24
30	36	33	37	31	34
36	37	38	37	41	41
47	45	49	45	48	47

Page 34

48

79

84

82

80
7
87

14
70
84

13
3, 1
9
93

12
2, 1
9
92

83	60	67	98	98	89
47	85	59	56	99	76
81	72	98	83	54	81
91	97	43	85	30	85
112	122	114	115	131	104
153	145	111	163	102	107
72	38	72	93	127	132
47	98	83	73	112	113

Page 36

75, 105, 114

20
18
200

28
8
2
18
188

101	112	112	135	113	96
127	107	125	148	125	118
72	137	114	126	125	109
141	138	138	130	138	131
157	167	138	150	152	170
133	142	151	157	212	190

Page 38

165		144
122	209	128
12,365	117	12,944
	11,909	

28	22	17
22	13	13
10	11	24
12	18	14
13,248	19,252	16,547

20,039 30,431 20,987

2,392	2,601	1,748	2,327	1,565
2,390	1,905	2,364	2,765	2,183
21,888	25,170	22,290	24,194	20,170
21,872	24,819	22,332	24,271	20,286

1,522,517 185,628,692

Page 40

170	280	180
240	440	210
670	750	200
1,350	2,730	
		1,800
700	300	2,000
600	800	2,022
5,200	2,600	
5,700	4,700	21,000
		24,000
3,000	313,000	23,313
9,000	859,000	
25,000	4,628,000	210,000
57,000	8,842,000	240,000
		238,000
Ten		238,536
Hundred		2,000,000
		2,200,000
Thousand		2,220,000
		2,226,161
Thousand		
		21,000,000
Hundred		24,000,000
		23,700,000
Ten		23,688,139
Ten		
Hundred		
Thousand		
Ten thousand		
Hundred thousand		

Page 41 Achievement Test

17	14	16	33	25	27	26
65	97	96	200	244	232	207
270	240	290	800	2,080	2,021	1,317
24,138	1,790	28,729	1,411	2,133,921	251,931	

329
275
253
211
238
280

264 344 375 380 223 = 1,586

Page 42 Related Problems

$7.12 $4.65 $7.25

$19.02

$8.62

$12.98

$40.62

$202.95 $436.35

Page 43 Related Problems

$192.32
153.61
160.81
140.59
140.96
168.40
200.49

$234.91 $227.00 $218.74 $227.82 $248.71 = $1,157.18

293,979 $1,626,615.47 250,488,000

ANSWER SHEET–Chapter 2

Page 48 Pretest

21 28 35 25 53 40 356 530 53

759 2,920 4,270 3,109 5,779

3,198 1,916 3,114 2,327 3,948

169,568 734 10,239 18,998

297,399 3,007

Page 58

5	6	8	4	3	7	9	6	8	8
9	3	2	1	5	10	2	0	7	5
4	5	5	7	7	2	5	8	5	3
6	7	7	8	4	6	3	8	10	10
4	2	8	7	3	7	4	9	4	11
0	8	7	6	0	6	0	4	8	6
5	4	4	6	3	9	3	8	12	2
14	11	14	8	6	4	7	8	9	11
6	5	10	6	6	6	11	14	17	16

Page 58

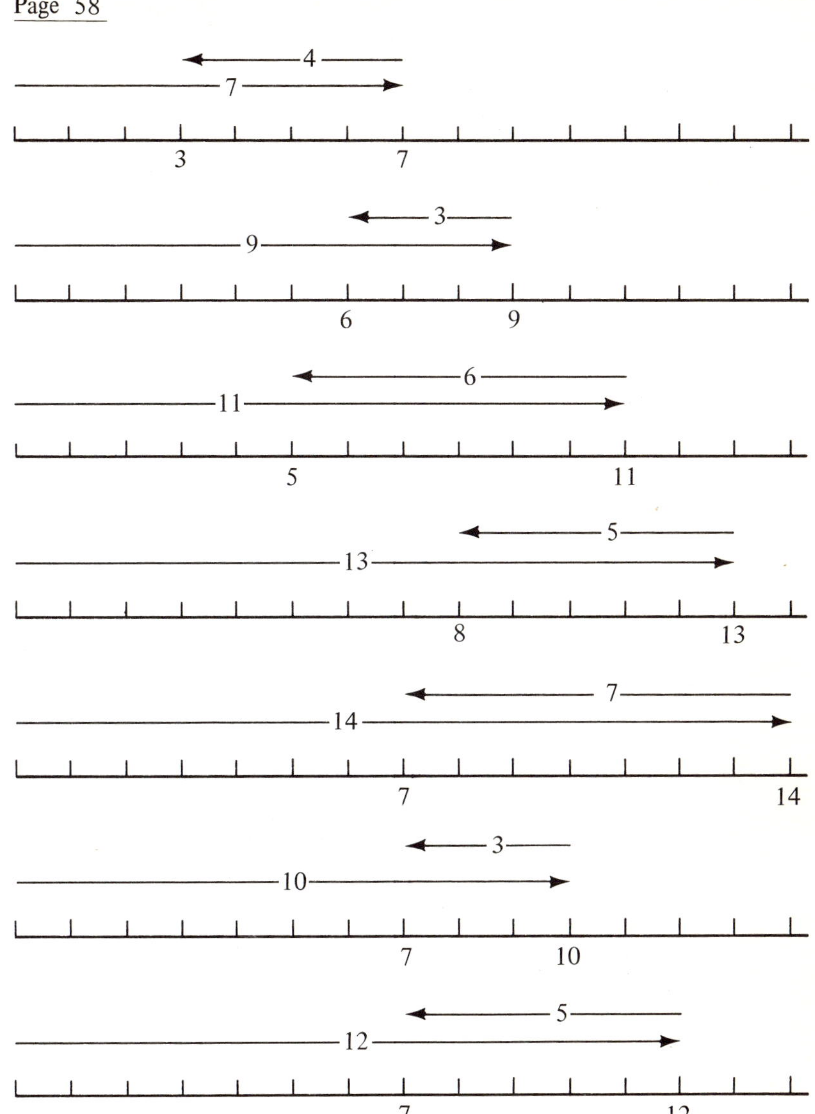

ANSWER SHEET–Chapter 2

Page 60

17	23	34	45	56	67	76	85	94
15	23	34	45	56	65	74	83	92
13	23	34	45	54	63	72	81	90
11	23	34	43	52	61	70	81	90
9	23	32	41	50	61	70	81	90
4	18	27	36	45	54	63	74	85
12	24	33	42	51	60	71	82	93
14	24	33	42	51	62	73	84	95
13	21	30	41	50	61	72	83	94
13	21	30	41	50	61	72	83	94

Page 62

47	14	49	17	17	29	77	56	47	33
79	88	38	79	59	59	88	19	29	46
47	29	74	47	88	69	57	65	29	74
37	12	26	61	19	76	38	88	19	89
69	63	18	39	68	87	66	86	39	67
68	58	87	43	35	47	19	25	27	57
26	37	84	56	36	78	64	57	76	48
38	78	66	55	44	19	68	64	56	13
59	27	24	28	43	49	36	25	58	64
74	59	55	85	22	85	88	32	18	68

Page 64

43	54	21	35	64	11	23	23
13	61	11	25	12	16	13	18
63	35	21	22	43	13	21	24
45	41	51	16	37	32	51	12
13	44	24	23	53	57	14	64
23	28	47	22	23	33	38	16
23	32	55	56	29	24	23	26

Page 66

37	46	19	25	56	9	17	17
7	59	9	15	8	4	7	2
57	25	19	18	37	7	19	16
35	39	49	4	23	28	49	8
7	36	16	17	47	43	6	56
17	12	33	18	17	27	22	4
17	28	45	44	11	16	17	14

ANSWER SHEET–Chapter 2

Page 68

123	245	413	522	332	732	231	632
512	424	336	435	652	923	861	761
237	348	515	419	817	528	617	817
547	636	828	925	757	736	617	507
192	884	472	683	761	433	852	793
793	571	786	865	772	682	585	661
588	878	467	669	788	667	868	788
778	569	776	388	779	447	884	468

Page 70

222	313	221	821	511	623	312	321
444	114	314	220	215	271	132	171
217	307	217	815	507	618	329	318
547	168	327	218	228	269	126	165
182	281	149	764	451	587	474	483
482	292	281	183	572	162	254	364
146	285	188	683	486	177	288	277
476	266	487	586	477	589	349	278

Page 72

30	20	20	10	20	60	20	0
33	26	27	13	28	62	25	8
405	501	603	303	802	903	700	202
395	499	597	297	798	897	698	198
450	510	630	330	820	930	700	220
350	490	570	270	780	870	650	180
444	504	626	328	813	921	692	211
123	806	631	324	511	716	412	523

Page 74

170	260	380	440	520	650	730	860
169	253	374	433	515	646	727	852
276	452	543	325	162	874	663	736
268	448	536	317	153	868	656	727
280	470	570	370	180	880	690	790
277	466	563	362	171	877	683	781
297	395	194	299	94	197	195	197

1,055 1,950 1,994 2,945 100,120 203,297 1,915,904

ANSWER SHEET—Chapter 2

Page 76

400	300
400	270
408	262
400	800
380	730
373	731
3,000	2,000
2,800	2,700
2,796	2,656
6,000	4,000
6,700	4,000
6,615	4,042
30,000	20,000
26,000	23,000
26,263	23,159
100,000	3,000,000
190,000	3,500,000
189,794	3,454,888

Page 77 Achievement Test

28	76	18	48	22
242	78	414	317	163
951	936	5,437	1,561	2,983
26,473	45,983	40,365	31,129	15,019

33
342
3,844
257
2,341

Page 78 Refresher Test

427	378	455	322	426
42,867	28,883	30,149	35,713	35,729

Page 79 Related Problems

$676.31	$609.51
$661.12	$465.68
$522.74	$432.91
$479.87	$143.38
$270.28	$ 91.35
$697.87	$419.80
$609.51	$410.06
	$292.28

Page 80 Related Problems

$25,287.50

34.50

547.25

$421.85

$35.06

7,555 miles

$171.84

ANSWER SHEET–Chapter 3

Page 84 Pretest

1,090	6,664	6,864	4,525	7,650
35,560	48,072	45,600	3,480	3,430
3,010	7,200	40,420	35,373	32,200
306,432	750,400	266,560	62,650	207,200
320,040	538,200	234,955	2,895,720	4,956,000

Page 96

15	14	6	16	20	21	15	35	12	25
48	36	24	12	16	16	30	6	28	6
24	63	42	56	5	24	0	4	18	30
12	4	42	24	54	0	9	40	12	0
10	24	15	35	45	49	32	8	20	12
54	21	8	0	32	64	40	15	0	45
45	16	27	9	20	14	28	63	27	8
72	36	24	10	0	18	9	36	56	40

ANSWER SHEET–Chapter 3

Page 98

84	246	160	129	124	279	288	39
88	155	126	146	100	186	128	105
279	208	287	240	486	200	108	48
186	328	300	357	320	279	208	126
147	208	400	355	540	156	126	70
168	249	350	427	640	148	549	210
69	355	450	328	497	240	567	630
155	450	147	189	248	128	280	540

Page 100

96	255	188	132	152	285	296	48
96	165	144	156	125	204	134	130
288	232	329	264	510	260	118	68
216	340	320	371	336	288	228	138
189	224	440	370	576	177	168	329
192	261	380	455	672	158	558	231
84	385	477	344	525	272	588	666
190	486	175	201	264	132	315	558

Page 102

1,083	1,625	772	1,686	872	4,291
1,170	1,938	1,278	1,952	1,099	2,292
2,316	1,317	8,262	3,766	3,072	6,760
3,199	2,316	2,394	1,972	3,468	2,445
2,016	2,916	2,344	4,670	3,078	2,556
1,072	4,110	4,260	4,232	2,655	2,544
5,256	2,292	4,434	1,980	1,928	2,580

Page 104

4,032	3,530	818	1,449	3,654	2,745
1,228	4,563	2,418	3,612	4,540	5,616
2,580	4,830	1,750	1,960	3,240	1,080
3,850	6,880	4,340	3,920	2,150	2,310
24,054	56,048	20,016	32,012	63,035	54,072
38,472	67,228	54,440	27,535	43,848	32,828
30,080	27,405	56,434	40,432	16,192	54,222
27,240	16,400	64,240	30,360	28,160	15,450

ANSWER SHEET—Chapter 3

Page 106

1,457	2,232	792	532	2,331	1,872	1,564	2,144
663	1,134	2,296	3,087	2,700	4,617	3,136	1,085
2,256	2,584	972	3,196	2,001	1,554	1,276	4,030
4,824	846	1,456	3,034	777	3,268	6,768	1,012
2,368	3,652	2,208	3,712	1,827	1,729	912	2,392
1,704	4,466	1,692	2,856	2,050	2,537	2,496	1,598

Page 108

12,082	10,925	27,432	21,199	40,865
12,122	9,369	3,304	15,648	52,472
14,706	12,544	36,153	19,440	9,882
58,632	339,855	81,168	117,777	164,348
76,755	480,524	390,488	60,625	365,472

Page 110

47,502	71,136	94,770	118,404	142,038	165,672	189,306
137,712	183,768	229,824	275,880	321,936	367,992	414,048
46,920	69,360	91,800	114,240	136,680	159,120	181,560
129,920	174,580	219,240	263,900	308,560	353,220	397,880

ANSWER SHEET—Chapter 3

Page 112

89,984	172,480	160,855	316,820	452,166
347,700	226,250	145,600	207,060	315,900
107,143	201,637	566,938	491,397	781,014
98,040	296,800	241,020	204,680	438,480
93,936	237,180	427,130	370,840	546,920

Page 114

130	60
380	100
730	300
1,260	1,200
100	1,200
400	8,000
700	12,000
1,300	160,000
412,000	25,000
848,000	30,000
3,517,000	150,000
7,633,000	200,000

Page 114 (cont.)

ten hundred	1,000,000
thousand thousand	1,500,000
hundred ten	25,000,000
ten hundred thousand	40,000,000

Page 116

3,478 3,600	4,012 3,600	6,496 6,000	20,944 21,000
48,224 45,000	335,325 320,000	704,954 720,000	203,048 210,000
1,377,621 1,400,000	562,790 600,000	15,210,206 14,000,000	16,865,280 16,000,000
2,477,436 2,400,000	1,095,511 1,200,000	1,808,324 2,000,000	15,593,096 16,000,000
101,987,462 100,000,000	495,371,868 480,000,000	1,560,762,710 1,500,000,000	1,022,344,670 900,000,000

ANSWER SHEET—Chapter 3

Page 117 Achievement Test

48	63	161	296	196
832	2,016	7,990	4,248	1,672
2,850	28,350	20,240	4,140	23,157
23,026	170,235	85,362	322,185	125,563
25,314	46,280	54,810	25,040	302,666

Page 118 Refresher Test

1,921	1,201	22,030	263,291	969,940
16,132	14,804	5,374	555	
357	425	393	136	827
3,930	2,280	14,937	335,824	1,672,135

Page 119 Related Problems

11,160,000 miles in 1 minute
669,600,000 miles in 1 hour
16,070,400,000 miles in 1 day
5,865,696,000,000 miles in 1 year

340,020 miles
126,720 feet
78,720 feet
566,280 square feet
1,116 pounds
20,736 cubic inches
405 cubic feet

Page 120 Related Problems

$50,580.00
$10,324.80
$14,965.20
$25,290.00

$994.75

$1,614.00

$14,300.00

$2,477.28
$1,236.04
$878.28
$92.04
$676.00
$8,940.36
$2,860.00
174.20
141.44
793.00
790.40
624.00
1,686.36
825.24
1,040.00
$8,934.64

Balance $5.72

Page 124 Pretest

31	620	3,003	73
78 R1	781 R2	87 R2	666 R6
70	408	1,004	1,067
72	406	790 R19	78 R6
42 R30	194 R22	177 R3	1,015 R12
1,060	4,090	5,009	7,007
63 R330	66 R412	831 R121	

Page 134

4	7
3	10
2	40
6	6
4	7
3	8
2	5
	6
12	9
8	5
6	9
4	12
3	3
2	15
	4
	15
	12

Page 136

6	6	3	3	8	7
8	9	9	6	9	6
6	6	4	7	8	9
9	0	6	6	6	9
8	0	4	0	9	7
4	*	8	5	0	9
0	8	7	*	9	6
7	6	5	7	9	*

* unworkable

Page 138

3 R2	5 R2	7 R1	9 R2	10 R2	12 R2	13 R2	14 R2
3 R1	4 R2	5 R1	6 R1	7 R3	9 R2	10 R2	12 R1
3 R1	2 R3	4 R3	5 R3	6 R3	7 R4	11 R2	12 R4
3 R1	2 R3	3 R4	4 R3	5 R2	6 R1	8 R4	11 R1
2 R3	2 R6	3 R3	4 R1	4 R6	6 R1	8 R1	8 R6
2 R2	2 R4	2 R7	3 R2	4 R4	5 R1	6 R5	7 R5
2 R1	1 R5	2 R3	3 R1	4 R2	5 R3	6 R1	7 R7

ANSWER SHEET—Chapter 4

Page 140

11 R7	12 R5	16 R1	11 R4	12 R1	13 R1
26 R1	17 R3	12 R7	14 R6	15 R4	17 R2
12 R2	11 R4	21 R1	12 R5	18 R5	23 R1
25 R1	22 R2	16 R4	12 R6	12 R4	11 R1
17 R3	13 R4	14 R1	16 R1	17 R1	13 R5
14 R4	23 R2	16 R2	21 R3	32 R2	46 R2
22 R5	23 R1	37 R2	29 R2	16 R1	15 R4

Page 142

112	123 R1	134	219 R1	331 R1	425 R1
112	121 R1	132 R1	215	224 R2	312 R2
112	121 R3	122 R3	203	214 R1	222 R1
102 R1	114 R2	111	107 R2	113	118 R3
102	106	108	111	112	115 R4
102 R1	107	109	111	111 R4	114
102 R1	104	106 R3	108	111	112 R3

Page 144

16	35	43	44	32	81 R3	53 R4
22	18	16	44	33	80 R6	84 R3
66	98	45 R1	88 R2	12	92 R1	80 R2
54	76	41	42	46	31	83 R2
26	18	28	15	25	68 R3	37 R3
45	55	36	66	14	54 R4	45 R2

Page 146

500	300	400	500
200	1,000	2,000	1,000
200	2,000	500	2,000
3,000	9,000	6,000	20,000
200	900	400	100
80	60	700	700
200	100	900	3,000

ANSWER SHEET—Chapter 4

Page 148

517	314	212	611
428 R17	921 R20	727 R33	522 R52
83	53	93	73
8,126 R27	5,126 R13	3,128 R36	7,122 R71
3,142	9,123	6,182	7,253

Page 150

105	207	808	504
606	902	308	709
150	270	80	540
660	730	870	980

Page 152

1,700	3,200	4,600	2,800
5,800	6,200	8,900	9,200
1,060	3,070	4,090	7,020
1,006	3,007	5,009	7,007

Page 154

191 R126	811 R804	395 R299	105 R26
82 R455	63 R330	766 R785	627 R669
57 R748	2,504 R141	518 R399	554 R8
521 R560	66 R412	318 R333	3,188 R165
253 R1	122 R637	831 R121	3,410 R89

Page 155 Achievement Test

201	35	198	70 R5
875	18	23 R10	44
55	87 R40	12,204 R31	14,911 R17
8,041 R33	4,990 R53	7,319 R45	78 R2
64 R3	3,080	4,205	2,069 R6

Page 156 Refresher Test

3,401	2,505	370	10,032	899,707
59,347	188,599	6,248	924	59,347
503,810	193,515	2,138,136	2,387,640	3,062,448

Page 157

Related Problems

52,080 miles

2,170 miles per hour

$1,080

13 cubic yards

45 square yards

45 yards

28 miles

155 kilometers

96 miles

105 cubic feet

1,206 pounds

Page 158

Related Problems

$108

$42.77
$6.11

$18.00

72 yards

105 miles

118 miles per hour

$78 per month
$18 per week

180 gallons

315 miles

7 days

ANSWER SHEET–Chapter 5

Page 162 **Pretest**

1 1/8	2 4/15	3 7/8
3/8	1 1/15	7/8
9/32	1	3 9/16
2	2 7/9	1 7/12
1/4	5/13	5/16
20	15	
25	56	

Page 166

10
1/10

1/10		
2/10	1/5	
3/10		
4/10	2/5	
5/10	1/2	
6/10	3/5	
7/10		
8/10	4/5	
9/10		
10/10	5/5	1

equivalent
same value
2/5
3/5
4/5
5/5
10/10 = 1
5/5 = 1

6
1/6
2/6
4/6
6/6 = 1
1

Page 164

4	2/6
1/4	6/6; 1
3/4	4
1/4	6
4/4; 1	6
3	1/6
4	4
4	
	12
1/4	
	1/12
3	
	7/12
6	5/12
1/6	12/12; 1
4/6	7
	12
	12
	1/12
	7
	3
	5
	2
	5
	7

ANSWER SHEET–Chapter 5

Page 168

4	16	16	12	9
9	15	15	8	12
8	15	12	64	9
25	6	80	16	20
16	96	4	32	15
100	12	24	80	8

Page 170

3/16	3/8	1/4	1/3	2/3
1/10	2/5	5/6	1/3	1/5
3/4	3/4	4/7	2/5	5/12
5/6	1/6	5/6	3/10	15/16
5/8	7/16	7/12	1/10	2/5

Page 172

2 6/7	9 1/2	9 1/3	3 2/3	2 3/5
3 4/5	10 3/4	4	6	3 3/16
3 9/32	14 1/3	4 1/2	2 9/32	5 7/16
10 2/5	2 1/2	8	4 7/15	3 1/12
4 3/8	5 1/7	3 5/12	6 1/8	5 5/7
9 3/5	4 3/11	6 7/11	3 1/9	33 1/3

Page 174

24	10	24	36	49
96	75	120	66	26
96	91	135	120	125

24/5	53/12	59/18	51/4
115/16	61/16	163/40	57/2
59/7	28/3	79/7	100/3
61/6	133/6	107/8	93/13
49/9	53/20	103/50	227/16

Page 176

3/4	1	1 2/5	1 2/3	1 2/9	1 1/2
1 1/4	1 1/7	1 1/5	1	1	1 1/2
1 1/2	1	1 1/5	1/2	8/9	1 1/5
1 7/8	1 5/9	1 3/5	2	1 6/7	1 9/10

ANSWER SHEET—Chapter 5

Page 178

1 7/9	4 2/3	3 5/7	5 7/8	2 3/4
5 3/5	6 5/7	8 1/2	8 2/3	11 1/2
6	7	7	5	10
8 1/3	9 3/10	11 1/4	9 1/3	5 1/2
8 1/8	12 1/16	9	11 1/3	9 1/6

Page 180

5/6	3/4	7/10	2/3
9/14	3/5	5/8	11/18
7/12	8/15	1/2	10/21
9/20	5/12	11/28	3/8
1 3/20	1 4/15	1 7/12	1 19/30
1 1/2	13/14	1 5/28	1 2/21
1 1/35	1 11/42	1 1/8	1 7/24
1 3/8	1 17/40	1 11/24	1 1/9
1 7/36	1 11/45	1 5/18	1 19/63

Page 182

2 3 5 7
11 13
17 19
23 29
31
37 41 43
47

2 . 2 . 2 . 2

2 . 3 . 3

2 . 2 . 5

3 . 7

2 . 11

5 . 5

2 . 13

3 . 3 . 3

2 . 3 . 5

10, 15, 20, 25, 30, 35, 40, 45, 50
14, 21, 28, 35, 42, 49, 56, 63, 70
18, 27, 36, 45, 54, 63, 72, 81, 90
24, 36, 48, 60, 72, 84, 96, 108, 120

2, 3, 4, 6
3, 5
2, 3, 6, 9
2, 4, 5, 10
3, 7
2, 3, 4, 6, 8, 12

2 . 2 . 2 . 3 2 . 2 . 2 . 7 2 . 2 . 3 . 3 2 . 2 . 2 . 2 . 3 3 . 3 . 3 . 3

Page 184

1 1/12	1 3/16	1 8/15	35/36
31/40	1 1/24	1 17/24	1 1/3
173/210	1 1/16	17/32	11/20
25/28	53/60	1 21/40	1 17/36
43 /72	91/120	1 1/48	1 13/36

Page 186

23/30	1 13/24	1 5/18
1 7/12	1 13/24	1 2/7
1 7/80	1 43/72	2 13/60
1 13/48	1 37/48	2 1/8

3 13/15	45/64
2 5/9	2 26/45
1 25/36	1 29/45

Page 188

1/2	1/7	0	1/3	3/5
1/2	1 1/7	2 1/5	3 1/3	4 2/9
5 1/4	4 2/7	7 2/5	6 1/6	11 2/9
4 3/4	3 5/7	6 3/5	5 5/6	10 7/9
4 3/7	7 4/5	2 2/3	5 5/8	10 8/9

Page 190

1/2	1/3	7/16	7/15
2 3/8	3 3/4	3 3/8	4 1/2
3 3/10	4 13/36	5 5/24	3 17/28
6 31/40	7 7/24	12 16/21	18 3/4

Page 192

2 6/7	1 1/3	6 2/5	4 1/6
5 1/4	4 1/11	10	10 1/2
4 4/9	7 1/5	11 3/7	8 8/9
8 2/11	9 3/13	11 2/3	11 2/3
11 1/4	10 4/5	15 3/11	10 1/2

Page 194

10/21	26/45	9/20	2/5
6/11	3/8	22/51	19/24
5/28	7/18	3/5	27/85
7/24	4/7	7/20	25/52
12/25	5/9	7/16	11/28

Page 196

1 7/12	4 13/20	10 5/48	29 1/3
6 5/14	4 2/3	8 1/10	
78 2/3	79 2/7	122 2/5	91 7/8
117 1/2	134 2/3	101 3/5	
14 7/9	39 6/7	53 1/8	128 1/3
70	115 5/7	144	

Page 198

12 1/2	6	14	36
14 2/5	17 1/7	47 1/4	46 1/5
5	2	3	3
2 1/10	1 2/9	1 7/8	1 1/5
16/21	9/11	8/15	5/6

Page 200

7	19	17	34
5	7 1/8	8 1/2	10 1/5
18	18 4/5	50	24 5/6
2/9	4/25	5/24	7/16
3/32	1/12	1/30	1/12

Page 202

2 17/32	6 1/14	5 3/5	2 5/8
4 5/8	2 11/56	3 29/36	2 1/3
1 13/21	2 1/4	2 1/2	6 2/3
21/34	4/9	2/5	3/20
6 38/45	5 1/3	5 23/45	4 26/49

Page 204

4/81	8/29	8/27	2/105
7/150	3/65	7/20	6/47
2/105	1/34	12/125	2/11
6 2/13	4 4/11	4 1/6	2 42/43
2 10/21	6 2/9	1 3/5	2 2/9
3 3/17	4 8/13	6 3/25	3 3/4

Page 206

1/9	1/8	2/5	1/2
30	56	60	140
96	378	128	27
98	48	3 3/4	73 1/2
475 1/5		120	
71 7/13		135	

Page 208

2 13/36	1 37/140	2 122/135	1 43/63
40/81	72/245	140/387	70/99
13/50	1 347/360	1 7/80	502/651
35/48	27/196	49/150	15/64
2 29/80	27/64	32/75	4/15

Page 210

21/52	14/39	15/61	15/53
3 20/21	3 5/12	4 12/35	2 22/35
5 5/18	4 7/78	2 106/177	
4 2/7	1 7/110	4 13/30	

Page 211 Achievement Test

6 3/5	15 7/24	14 7/8	24 1/3
1 1/5	4/5	2 4/35	2 7/16
5/16	1/12	35	55
10 3/20	29	1/4	
2/3	1 1/2	1 3/5	
2/3	2 4/7	44/147	
3 9/16	17 2/3		

Page 212 Refresher Test

56	512	4,599	27,090	8,432
33	474	8,218	58,778	274,898
270	704	7,362	26,790	3,285,205
15 R5	8 R35	53 R25	707	3,009

Page 213

325 square inches
1,173 cubic inches
29 1/4 pounds
8 1/3 hours
67
12
50 cents
20 cents
$1.98

Page 214

17 5/8 inches
46 1/4 yards
2 15/32 inches
3 7/8 pounds
11 9/16
7/16
1/2
5/12 hour

Page 218 Pretest

		43.2018	
		10.022567	
		9.042	
		.757	
2.496	8.799	162.036	32,412.6
7.5	8.5215	2.52294	254.275
.05	1,200	1.1	.0025
150,000	.0004	7	.0071

Page 223

Answers to both columns:
two tenths
three hundredths
twenty-five hundredths
five thousandths
thirty-five thousandths
four hundred fifty-five thousandths
seven ten thousandths
forty-eight ten thousandths
one hundred fifty-five ten thousandths
three thousand five hundred twenty-five ten thousandths
three hundred thousandths
seventy-eight hundred thousandths
two hundred twenty-nine hundred thousandths
five thousand five hundred seventy-five hundred thousandths
sixty-two thousand one hundred twenty-three hundred thousandths
eight millionths
fifty-six millionths
two hundred five millionths
three thousand six hundred two millionths
forty-seven thousand one hundred thirty-five millionths
eight hundred twenty-five thousand two hundred twenty-seven millionths

Right column only:
two AND four hundredths
thirteen AND five thousandths

Page 224

.01
.05
.1
.15
.25
.65
.9
1.05
1.25
10.01
10.1
50.05
50.5
100.01
100.1

Page 226

49.7723
196.1665
571.363

10.52	12.35	10.43	11.87	13.87
17.78	6.77	6.68	9.55	2.086
11.397	89.25	39.06	2.064	24.11
.2655	.1851	.1889	.02199	.02227
85.04	125.7	93.92		

Page 228

4.87		2.27	3.29	
5.71		.48	7.08	
3.81	4.43	5.71	6.28	2.75
1.268	3.6083	6.392	79.205	4.803
.0338	.0787	.00050	.00194	.00135
63.04	34.12	81.82	28.9927	15.9912
.000867	.0000393	.0000694	.0000313	

ANSWER SHEET—Chapter 6

Page 230

.6	1.0	2.8	2.0	2.4	3.0	7.2	4.0
.12	.1	.32	.2	.18	.3	.56	.4
.88	1.1	2.64	2.2	3.96	3.3	2.64	4.4
1.54	1.1	1.76	2.2	4.62	3.3	5.28	4.4
.198	.11	.132	.22	.528	.33	.704	.44

Page 232

.216	.352	.558	.285	.546
.00252	.00264	.00496	.00513	.0039
.0000396	.0000572	.0000868	.0000855	.000195
	.06		.07	
	.12		.108	
	.00144		.0012	
	.0000144		.0000161	
	.000000157872		.0000009702	
	.0000002376		.00000019845	

Page 234

5	2.5	20	30
30	20	20	40
50	25	200	300
200	300	400	200
50	25	40	110
120	50	110	70
500	250	500	900
1,000	1,400	1,000	1,100

Page 236

.3	.2	.3	3.1	2.1	3.2
.04	.12	.11	.4	1.2	1.1
.125	.122	.112	1.25	1.22	1.12
.124	.223	.443	1.24	2.23	4.43
.1025	.1224	.1305	10.25	12.24	13.05
.10118	.33006	.23004	101.18	330.06	230.04

ANSWER SHEET—Chapter 6

Page 238

2	4	3	21	42	32
30	.8	1.65	3	8	11
1.45	1.44	1.22	14.5	14.4	12.2
1.56	2.77	3.76	15.6	27.7	37.6
1.235	1.409	1.008	123.5	140.9	100.8
1.0409	2.1305	2.3662	1,040.9	2,130.5	2,366.2

Page 240

5	6	6	7	.5	.5
60	70	2.8	.48	.75	.225
120	150	63	7.4	.08	.16
370	280	1.87	28.9	.048	.0128
2,240	3,280	23.05	300.6	.00064	.0075
40,150	30,080	28	62	.00008	.0008

Page 242

4/5	3/5	3/10	.5	.8	.75	23.3125
3/25	3/20	7/20	.4	.24	.25	18.75
3/8	21/125	133/200	.07	.065	.12	212.25
5/16	1/80	1,137/2,000	3.25	568.125	2.5	
4 1/2	6 4/5	15 3/10	12.9		5.125	
3 17/20	4 12/25	11 1/20	29.62		45.72	
12 1/8	15 3/40	28 9/125	125.5		42.185	

ANSWER SHEET—Chapter 6

Page 244

.4	.25	.7	.33	.6
.23	.78	.285	.34	.47

.024	.016	.015	.013
.043	.038	.041	.038
.042	.028	.043	.23

Page 246

.0034	.0061	670	2,500
.0016	.006	480	420
.0036	.0081	850	370
.00016	.0063	280	240
.009175	.0046	620	350

Page 248

128	256	.67	4.9
16	64	.43	.42
64	32	.24	2.7
256	32	.23	.039
69.12	3,437.5	.038	.047

Page 250

.333	.833	.444	.6363	.4545
5.833	7.77	4.5454	1.666	4.166

.3	.05	.007	.0009	.00004
.000006	.15	.023	.0037	.00063
.000045	.115	.0234	.00513	.000718
.5275	.02871	.008329	.32753	
.072561	.419236			

.6	.5	.4	.6	.3	2.4	15.7
.62	.49	.28	.45	.83	2.44	15.69
.625	.489	.835	.569	2.437	15.694	

Page 252

94	7	.91	.12
1.4	.0225	.044	1,000
227	.153	2,600	.03
11200	.9125	.06	.00445

ANSWER SHEET–Chapter 6

Page 254

274	4,767	17.16	129.95
.00528	.0875	.0001875	.00018
701	.001335	12,003.034	
120,000	.75	.1234	

Page 255 Achievement Test

		2.697	.49729	70.92
		14.87		
.22	2.39	16.705	.000886	
3.04	.00588	1.256	.6954	
.0004	.0003		.0005738	
380	340		54,700	

Page 256 Refresher Test

2,862	2,477,867	3.3627	6.722	10 1/8	15 23/24
7,477	736,396	.00436	11.0875	3 1/8	7 13/24
34,611	5,713,658	.61425	.0000217	21	56
468	3,807 R9	25,000	.000015		
2 2/17	2 8/25				

Page 257 Related Problems

4.86 square feet

252.252 cubic feet

32

19.5 pounds

2.1 dollars

15

$1.12

$ .48
$2.52

Page 258 Related Problems

4.35 feet

36.95 yards

1.635 inches

3.35 pounds

10.7 feet

1.3 feet

.44

24 hours

ANSWER SHEET—Chapter 7

Page 262 Pretest

5%	28%	120%	.2%
.16	.03	.002	1.25
3/20	3/8	2/3	1/4
7.2	60	15%	
14.95	64	8%	
.63	500	.5%	

Page 264

hundred
10
100
%
percent
22 100
twenty-two
100

2/100	12/100	26/100	31/100	43/100
55/100	64/100	87/100	98/100	100/100
3%	11%	23%	35%	
42%	52%	74%	100%	

Page 266

50%	
25%	
10%	
5%	
1%	
50%	8%
50%	
50%	35%
50%	
5%	75%
10%	
20%	100%
25%	
50%	
100%	

Page 268

.05	.02	.01	.1	.5
.25	.15	.75	.0375	.0175
.0535	.0845	1.	2.5	7.15
3.32	.375	.833	.872	.15125
.23	.66	.03	.18	.27
.164	.257	.462	.123	.338
.0047	.0063	.0015	.0098	.001
.005	.0025	.002	.00125	.001

ANSWER SHEET–Chapter 7

Page 270

5%	10%	1%	50%	2.5%
25%	.5%	20.5%	38.5%	1.5%
73.5%	.8%	200%	500%	250%
125%	305%	165%	442%	625.5%
3.35%	8.62%	5.45%	.95%	62.75%
33.45%	22.62%	76.48%	12.5%	37.5%
62.5%	87.5%	.23%	.76%	.14%
.99%	105.25%	400.36%	202.37%	807.25%

Page 272

20%	50%	90%	100%	200%
40%	70%	30%	150%	300%
2%	15%	70%	150%	300%
7%	5%	25%	40%	100%
2.2%	1.5%	7%	15%	30%
.5%	2.5%	25%	70%	100%

Page 274

50%	33 1/3%	25%	20%	16.7%
14.3%	12.5%	11.1%	66 2/3%	75%
60%	42.9%	80%	57.2%	44.4%
37.5%	83.5%	71.4%	62.5%	85.7%
18.75%	9.375%	4.6875%	31.25%	15.625%
72%	85%	38%	56.7%	48.9%

Page 276

1/100	1/50	3/100	1/25	1/20
3/50	7/100	2/25	9/100	1/10
3/20	1/5	1/4	3/10	7/20
2/5	9/20	1/2	11/20	3/5
13/20	7/10	3/4	4/5	17/20
9/10	19/20	1	1 1/4	1 1/2
1 3/4	2	2 1/2	3	5
1/8	3/8	5/8	7/8	1/3
2/3	1/200	1/500	1/1,000	1/4,000

ANSWER SHEET–Chapter 7

Page 278

56	128	6,240
28	64	3,120
14	32	1,560
5.6	12.8	624
2.8	6.4	312
.56	1.28	6.24
.056	.128	.624

.72	2.1
26.46	2.24
1.25	8
32	45
360	364.5
10.2	31.96
.33	5.5

Page 280

1,000%	1,000%
100%	100%
50%	50%
25%	25%
10%	10%
5%	5%
1%	1%
.1%	.1%
25%	20%
10%	10%
50%	50%
1%	5%
5%	1%
33 1/3%	33 1/3%
66 2/3%	420%
100%	15%
1,000%	22%
12.5%	1.5%

Page 282

20	4.5
200	45
400	90
800	180
2,000	450
4,000	900
20,000	4,500
200,000	45,000
200	187.5
75	96
448	40
100	20
260	80
17.5	37.5
3,500	163
600	305
75	850
180	150

Page 283 Achievement Test

.08	.15	.8	2.25
7%	55%	.2%	350%
1/5	7/8	1/3	3/4
11.04	32.5		
4.84	213		
300	80		
200	94.5		
25%	15%		
12.5%	5%		

Page 284 Refresher Test

202	173.96	22.452	88,559
239	52.3	44.45	164,404
15.75	23,374	72.384	1,537,994
.6625	.029	4,890 R5	
89 R15	7,100	6,862	

Page 285	Related Problems
$315	
1,080	
$1.25	
$396	
.64 ounces	
31.36 ounces	
1.6 ounces	
30.4 ounces	
$2,293.50	
156	

Page 286	Related Problems
$22,500	
$35,000	
37 1/2%	
62 1/2%	
31.9%	
6%	
66 2/3%	
53.6%	
46.4%	
13 1/3%	

ANSWER SHEET—General Mathematics Test

Page 289

35	585	13,163	347
414	2,877.99	10,728	595.28
386	1,445	3,439	14,066

Page 290

474,240	88	355,984	507
52,333	904	51,896	86 R18
525,600	74 R12	400,020	640 R34

Page 291

713,542	6.92 R40	1 9/28	1 4/15
36	1 1/11	5/6	3 7/16
3/8	129	1/6	2/3
2 3/4	1/25	1 3/7	.15

Page 292

.8	20.75	350	.13
.0325	30	1%	5
2.25	30%	2 1/2%	
406.4	$6,000	520